Advances in Engineering Materials

Preparation for Sustainable Process Development

Advances in Engineering Materials

Preparation for Sustainable Process Development

Tamara Tatrishvili, PhD

A. K. Haghi, PhD

Editors

AAP | APPLE ACADEMIC PRESS

First edition published 2025

Apple Academic Press Inc.
1265 Goldenrod Circle, NE,
Palm Bay, FL 32905 USA

760 Laurentian Drive, Unit 19,
Burlington, ON L7N 0A4, CANADA

CRC Press
2385 NW Executive Center Drive,
Suite 320, Boca Raton FL 33431

4 Park Square, Milton Park,
Abingdon, Oxon, OX14 4RN UK

Library and Archives Canada Cataloguing in Publication

CIP data on file with Canada Library and Archives

Library of Congress Cataloging-in-Publication Data

CIP data on file with US Library of Congress

ISBN: 978-1-77964-372-8 (hbk)
ISBN: 978-1-77952-037-1 (pbk)
ISBN: 978-1-00363-804-9 (ebk)

About the Editors

Tamara Tatrishvili, PhD

Assistant Professor, Main Specialist, Office of the Academic Process Management, Ivane Javakhishvili Tbilisi State University (TSU); Director of the Institute of Macromolecular Chemistry and Polymeric Materials at TSU, Tbilisi, Georgia

Tamara Tatrishvili, PhD, is an Assistant Professor and Main Specialist at the Office of Academic Process Management (Faculty of Exact and Natural Sciences) at Ivane Javakhishvili Tbilisi State University as well as the Director of the Institute of Macromolecular Chemistry and Polymeric Materials at TSU, Tbilisi, Georgia. She is also a DAAD alumna and a member of the Georgian Chemical Society. Her research interests included polymer chemistry, polymeric materials, and chemistry of silicon-organic compounds. She is the executive editor of the *Journal of the Georgian Chemical Society.* Dr. Tatrishvili is the author of more than 190 scientific publications, 12 books, and several monographs.

A. K. Haghi, PhD

Research Associate, Department of Chemistry, University of Coimbra, Portugal

A. K. Haghi, PhD, is a retired professor and has written, co-written, edited, or co-edited more than 1000 publications, including books, book chapters, and papers in refereed journals with over 3800 citations and an h-index of 32, according to the Google Scholar database. He is currently a research associate at the University of Coimbra, Portugal. Professor Haghi has received several grants, consulted for several major corporations, and is a frequent speaker to national and international audiences. He is the Founder and former Editor-in-Chief of the *International Journal of Chemoinformatics and Chemical Engineering* and *Polymers Research Journal.* Professor Haghi has acted as an editorial board member of many international journals. He has served as a member of the Canadian Research and Development Center of Sciences & Cultures. He has supervised several PhD and MSc theses at the University of Guilan (UG) and co-supervised international doctoral projects. Professor Haghi holds a BSc degree in urban and environmental engineering from the University of North Carolina (USA) and holds two MSc, one in mechanical engineering from North Carolina State University (USA) and another one in applied mechanics, acoustics, and materials from the Université de Technologie de Compiègne (France). He was awarded a PhD in engineering sciences at Université de *Franche-Comté* (France). He is a regular reviewer for leading international journals.

Contents

Contributors

Gunel Abaszade
Institute of Polymer Materials of the Ministry of Science and Education of the Republic of Azerbaijan, Sumgait, Azerbaijan

Rena Akhmedova
Institute of Polymer Materials of the Ministry of Science and Education of the Republic of Azerbaijan, Sumgait, Azerbaijan

Dilbar Aliyeva
Institute of Polymer Materials of the Ministry of Science and Education of the Republic of Azerbaijan, Sumgait, Azerbaijan

Rômulo Simões Angélica
Universidade Federal do Pará (UFPA), Belem, Brazil

Melina Babolhavaeji
Department of Chemistry, Faculty of Science, Imam Khomeini International University, Qazvin, Iran

Bendeliani Bezhan
Ilia Vekua Sukhumi Institute of Physics and Technology, Laboratory of Cryogenic Technique and Technologies, Mindeli St. 7, Tbilisi, Georgia

Nodar Beridze
Scientific-Research Institute of Organic Chemistry Faculty of Exact and Natural Sciences Iv. Javakhishvili Tbilisi State University, Tbilisi, Georgia

Rebeca Bouza
Polymer Group, Physics Department, University of A Coruña, Polytechnic University School, Ferrol Campus, Ferrol, Spain

Klaudia Brzezinska
Department of Biomaterials and Cosmetic Chemistry, Faculty of Chemistry, Nicolaus Copernicus University in Torun, Poland

Tinatin Bukia
V. Chavchanidze Institute of Cybernetics, Georgian Technical University, S. Euli st. 5, Tbilisi, Georgia
Scientific-Research Institute of Organic Chemistry Faculty of Exact and Natural Sciences IV. Javakhishvili Tbilisi State University, Tbilisi, Georgia
Institute of Macromolecular Chemistry and Polymeric Materials, Ivane Javakhishvili Tbilisi State University, I.Chavchavadze Ave, 13, Tbilisi, Georgia

Luka Buskandze
Scientific-Research Institute of Organic Chemistry Faculty of Exact and Natural Sciences IV. Javakhishvili Tbilisi State University, Tbilisi, Georgia

Efkan Çatıker
Department of Chemistry, Ordu University, Ordu, Turkey

Madlena Chkhaidze
Petre Melikishvili Institute of Physical and Organic Chemistry of the Iv. Javakhishvili Tbilisi State University, Tbilisi, Georgia

Ketevan Chubinidze
Institute of Cybernetics of Georgian Technical University, Z. Anjaparidze 5, Tbilisi, Georgia

Mariam Chubinidze
Jo Ann University Hospital, Ljubljana 35, Tbilisi, Georgia

Surmanidze Daviti
Ilia Vekua Sukhumi Institute of Physics and Technology, Laboratory of Cryogenic Technique and Technologies, Mindeli St. 7, Tbilisi, Georgia
Ivane Javakhishvili Tbilisi State University, Faculty of Exact and Natural Sciences, Department of Chemistry, Chavchavadze Ave. 3, Tbilisi, Georgia

D. S. Davtyan
Mckinven Expert LLC, Moscow, Russia

Maryam Dehghani
Universidade Federal do Pará (UFPA), Belem, Brazil

Aida Dunyamaliyeva
Institute of Polymer Materials of the Ministry of Science and Education of the Republic of Azerbaijan, Sumgait, Azerbaijan

Subrata Dutta
SKD Consulting, Glenn Dale, MD, USA

Abdollah Esmaeili
Universidade Federal do Pará (UFPA), Belem, Brazil

Alexander Gamkrelidze
Faculty of Exact and Natural Sciences, Tbilisi State University, Tbilisi, Georgia

Laura Garzón
Universidade Federal do Pará (UFPA), Belem, Brazil

N. K. Gasparya
National Polytechnic University of Armenia, Yerevan, RA

Shahriar Ghammamy
Department of Chemistry, Faculty of Science, Imam Khomeini International University, Qazvin, Iran

Ana Goletiani
Institute of Macromolecular Chemistry and Polymeric Materials, Ivane Javakhishvili Tbilisi State University, Tbilisi, Georgia

T. M. Guliyeva
Institute of Polymer Materials of the Ministry of Science and Education of the Republic of Azerbaijan, Sumgait, Azerbaijan

Dgebuadze Guram
Ilia Vekua Sukhumi Institute of Physics and Technology, Laboratory of Cryogenic Technique and Technologies, Tbilisi, Georgia

Manana Gurielidze
Durmishidze Institute of Biochemistry and Biotechnology, Agricultural University of Georgia, Tbilisi, Georgia

Shabnam Guseinova
Institute of Polymer Materials of the Ministry of Science and Education of the Republic of Azerbaijan, Sumgait, Azerbaijan

Zoia Haholkina
Polymer Group, Physics Department, University of A Coruña, Polytechnic University School, Ferrol Campus, Ferrol, Spain

Asgar Huseynov
Institute of Catalysis and Inorganic Chemistry Named After Academician M.Naghiyev, Ministry of Science and Education, Republic of Azerbaijan, Baku, Azerbaijan

Fedor Ilkov
IlkovF Consulting, Morgan Hill, CA, USA

Metskhvarishvili Ioseb
Ilia Vekua Sukhumi Institute of Physics and Technology, Laboratory of Cryogenic Technique and Technologies, Tbilisi, Georgia
Georgian Technical University, Faculty of Informatics and Control Systems, Department of Microprocessor and Measurement Systems, Tbilisi, Georgia

Ismail Ismailov
Institute of Polymer Materials of the Ministry of Science and Education of the Republic of Azerbaijan, Sumgait, Azerbaijan

Chimnaz Ismailova
Institute of Polymer Materials of the Ministry of Science and Education of the Republic of Azerbaijan, Sumgait, Azerbaijan

Zaza Jaliashvili
Department of Optics and Spectroscopy, Institute of Cybernetics, Georgian Technical University, Tbilisi, Georgia

Giorgi Kachlishvili
Department of Optics and Spectroscopy, Institute of Cybernetics, Georgian Technical University, Tbilisi, Georgia

Elene Kalandia
Institute of Cybernetics of Georgian Technical University, Tbilisi, Georgia

Temur Kantaria
Institute of Chemistry and Molecular Engineering, Agricultural University of Georgia, Tbilisi, Georgia

Tengiz Kantaria
Institute of Chemistry and Molecular Engineering, Agricultural University of Georgia, Tbilisi, Georgia

Kakha Karchkhadze
Petre Melikishvili Institute of Physical and Organic Chemistry of the Iv. Javakhishvili Tbilisi State University, Tbilisi, Georgia

Ramaz Katsarava
Institute of Chemistry and Molecular Engineering, Agricultural University of Georgia, Tbilisi, Georgia

A. G. Ketyan
National Polytechnic University of Armenia, Yerevan, Armenia
Department of Computer Science, Tbilisi State University, Tbilisi, Georgia

Oleg Kharshiladzre
Faculty of Exact and Natural Sciences, Tbilisi State University, Tbilisi, Georgia

Natela Khetsuriani
Petre Melikishvili Institute of Physical and Organic Chemistry of the Iv. Javakhishvili Tbilisi State University, Tbilisi, Georgia

Irine Khubua
Caucasus Medical Centre, Tbilisi, Georgia

Irakli Kirtadze
Jo Ann University Hospital, Tbilisi, Georgia

Irakli Koberidze
Department of Computer Science, Tbilisi State University, Tbilisi, Georgia

Taia Kotorashvili
Institute of Chemistry and Molecular Engineering, Agricultural University of Georgia, Tbilisi, Georgia

Kaloian Koynov
Max Planck Institute for Polymer Research, Mainz, Germany

Justyna Kozlowska
Department of Biomaterials and Cosmetic Chemistry, Faculty of Chemistry, Nicolaus Copernicus University in Torun, Poland

Nino Kupatadze
Institute of Chemistry and Molecular Engineering, Agricultural University of Georgia, Tbilisi, Georgia

N. I. Kurbanova
Institute of Polymer Materials of the Ministry of Science and Education of the Republic of Azerbaijan, Sumgait, Azerbaijan

Nushaba Kurbanova
Institute of Polymer Materials of the Ministry of Science and Education of the Republic of Azerbaijan, Sumgayit, Azerbaijan

Nazi Kutsiava
Department of Chemical and Biological Technologies, Georgian Technical University, Tbilisi, Georgia

Claudio Regis dos Santos Lucas
Universidade Federal do Pará (UFPA), Belem, Brazil

Metskhvarishvili Magda
"Talga" Institute of Georgian Technical University, Tbilisi, Georgia

David Makharadze
Departament d'Enginyeria Química, Universitat Politècnica de Catalunya, Barcelona, Spain

Tamar Makharadze
Institute of Cybernetics of Georgian Technical University, Tbilisi, Georgia

Bakhtiyar Mamedov
Institute of Polymer Materials of the Ministry of Science and Education of the Republic of Azerbaijan, Sumgait, Azerbaijan

G. H. Mamedova
Azerbaijan State Oil and Industry University, Azerbaijan

Samira Mammadova
G. M. Abdullayev Institute of Physics, Ministry of Science and Education, Republic of Azerbaijan, Baku, Azerbaijan
National Aviation Academy, Baku, Azerbaijan

Vako Marchilashvili
Department of Optics and Spectroscopy, Institute of Cybernetics, Georgian Technical University, Tbilisi, Georgia

Irina Mchedlishvili
Petre Melikishvili Institute of Physical and Organic Chemistry of the Iv. Javakhishvili Tbilisi State University, Tbilisi, Georgia

Tamaz Medoidze
Department of Optics and Spectroscopy, Institute of Cybernetics, Georgian Technical University, Tbilisi, Georgia

Nino Medzmariashvili
Department of Chemical and Biological Technologies, Georgian Technical University, Tbilisi, Georgia.

Zaza Melikishvili
Department of Optics and Spectroscopy, Institute of Cybernetics, Georgian Technical University, Tbilisi, Georgia

Ali Mirhashemi
Department of Chemistry, Faculty of Science, Imam Khomeini International University, Qazvin, Iran

Nurlana Mirzoeva
Institute of Polymer Materials of the Ministry of Science and Education of the Republic of Azerbaijan, Sumgayit

Afonso Cesar Rodrigues Nogueira
Universidade Federal do Pará (UFPA), Belem, Brazil

Dilbar Nurullayeva
Institute of Polymer Materials of the Ministry of Science and Education of the Republic of Azerbaijan, Sumgait, Azerbaijan

Natia Ochkhikidze
Institute of Chemistry and Molecular Engineering, Agricultural University of Georgia, Tbilisi, Georgia
Department of Chemical and Biological Technologies, Georgian Technical University, Tbilisi, Georgia.

Temel Öztürk
Department of Chemistry, Giresun University, Giresun, Turkey

Niloofar Palizdar
Department of Chemistry, Faculty of Science, Imam Khomeini International University, Qazvin, Iran

Ana Petriashvili
Department of Optics and Spectroscopy, Institute of Cybernetics, Georgian Technical University, Tbilisi, Georgia
Institute of Cybernetics of Georgian Technical University, Tbilisi, Georgia

Gia Petriashvili
Institute of Cybernetics of Georgian Technical University, Tbilisi, Georgia
V. Chavchanidze Institute of Cybernetics, Georgian Technical University, Tbilisi, Georgia

Sevinc Ragimova
Institute of Polymer Materials of the Ministry of Science and Education of the Republic of Azerbaijan, Sumgait, Azerbaijan

Farzane Yousefi Ramandi
Department of Chemistry, Faculty of Science, Imam Khomeini International University, Qazvin, Iran.

Maite Rico
Polymer Group, Physics Department, University of A Coruña, Polytechnic University School, Ferrol Campus, Ferrol, Spain

Julia Rypinska
Department of Biomaterials and Cosmetic Chemistry, Faculty of Chemistry, Nicolaus Copernicus University in Torun, Poland

Zaza Sopromadze
Jo Ann University Hospital, Tbilisi, Georgia

Nodar Sulashvili
Scientific Research-Skills Center, Tbilisi State Medical University, Tbilisi, Georgia
Faculty of Medicine, Sulkhan-Saba Orbeliani University, Tbilisi, Georgia

Tamara Tatrishvili
Institute of Macromolecular Chemistry and Polymeric Materials, Ivane Javakhishvili Tbilisi State University, Tbilisi, Georgia

Lobzhanidze Tea
Ivane Javakhishvili Tbilisi State University, Faculty of Exact and Natural Sciences, Department of Chemistry, Tbilisi, Georgia

A. O. Tonoyan
National Polytechnic University of Armenia, Yerevan, Armenia

Elza Topuria
Petre Melikishvili Institute of Physical and Organic Chemistry of the Iv. Javakhishvili Tbilisi State University, Tbilisi, Georgia

Magda Tsintsadze
Department of Computer Science, Tbilisi State University, Tbilisi, Georgia

Luka Tsulukidze
Faculty of Exact and Natural Sciences, Tbilisi State University, Tbilisi, Georgia

Gabunia Vakhtang
Petre Melikishvili Institute of Physical and Organic Chemistry of the Iv. Javakhishvili Tbilisi State University, Tbilisi, Georgia

A. Z. Varderesyan
National Polytechnic University of Armenia, Yerevan, Armenia

Mesut Yılmaz
Dept. of Chemistry, Ordu University, Ordu, Turkey

Nino Zavradashvili
Institute of Chemistry and Molecular Engineering, Agricultural University of Georgia, Tbilisi, Georgia

Eldar Zeynalov
Institute of Catalysis and Inorganic Chemistry Named After Academician M. Naghiyev, Ministry of Science and Education, Republic of Azerbaijan, Baku, Azerbaijan

Preface

The book covers the most recent developments in materials technology and science, providing basic as well as deep insight into advanced materials in regard to structural, physical behavior, and sustainable real-world applications. This research-oriented book reports on the most recent developments in the practical applications of novel materials for upcoming technologies attributed to their interesting and exciting mechanical and chemical behavior along with their numerous engineering aspects. It is shown that how these interesting engineering properties are responsible for material applications in different types of devices and sustainable sources of energy.

This new book also addresses the relationship between structural identities of materials with their property profile. It explains the way we control the process parameters during the preparation of materials in detail in order to obtain the desired features in the form of arrangement of atoms, ions, or molecules and thereby the preparation of the materials to be in-line with the real-world applications.

This book is intended for researchers, academicians, and scientists in the field of materials science and technology, computational physics, industrial technology, and so on.

CHAPTER 1

Investigation of Swelling Behavior of Polyepichlorohydrin Cross-Linked with Polyethyleneglycol Bridges in Solvents of Different Polarity

EFKAN ÇATIKER[1], MESUT YILMAZ[1], and TEMEL ÖZTÜRK[2]

[1]*Department of Chemistry, Ordu University, Ordu, Turkey*

[2]*Department of Chemistry, Giresun University, Giresun, Turkey*

ABSTRACT

Polyepichlorohydrin (PECH) has been cross-linked by the well-known alkyne-azide "click" reaction after it has been chemically modified. First, the modified PECH polymers (PECH25, PECH50, and PECH100) with different cross-linking potentials were synthesized by substitution of some of the pendant chlorine atoms in the PECH chains with azide groups at three different azidation rates (25, 50, and 100%). The determination of whether the synthesized partially azidified PECH polymers were azidified using Fourier transform infrared spectroscopy (FT-IR) spectroscopy and to what extent they were azidified was carried out using elemental analysis. In addition, the hydroxy end-groups of polyethylene glycols (PEGs), PEG-400 and PEG-1450, which were to be used as cross-linking agents (bridge), were converted into propargyl groups (PEG400-diP and PEG1450-diP). Then, the partially azidified PECHs were cross-linked via a "click" reaction using the dipropargyllated PEGS as cross-linkers, yielding various products (PECH25-PEG400, PECH50-PEG400, PECH100-PEG400,

Advances in Engineering Materials: Preparation for Sustainable Process Development. Tamara Tatrishvili and A. K. Haghi (Eds.)

PECH25-PEG1450, PECH50-PEG1450, and PECH100-PEG1450). The success of the "click" reactions has been confirmed by the disappearances of propargyl and azide bands in the FT-IR spectra of the products. The trace band at about 3132 cm^{-1} in FT-IR spectra belonging to the triazole group is also an indicator of success in alkyn-azide "click" reaction.

Mass proportions of PEG cross-linkers in the final products were achieved by thermogravimetric analysis (TGA). The swelling ratios of all cross-linked products in tetrahydrofuran (THF), *N,N*-dimethylformamide (DMF), methanol, and acetone were measured at 25 and 37°C. Maximum swelling degrees of all samples were found to be increasing with temperature. Similarly, in all solvents, the swelling degrees of the products with PEG1450 were higher than those with PEG400. For the samples with PEG400, it was observed that the degree of swelling decreased as the cross-linking ratio increased. However, no such regular change was observed for the samples with PEG1450. This anomaly is attributed to the change in the chemical properties of the cross-linked products due to the bulk dominance of PEG1450 in highly cross-linked products. Dependency of the degrees of swelling on the solvent type showed that DMF is the best solvent for all samples at both 25 and 37°C.

1.1 INTRODUCTION

Polyethylene glycol (PEG) is derived from petroleum for many applications such as cosmetics, medicine, pharmacy, etc. It is also known as polyethylene oxide and polyoxyethylene. Polyepichlorohydrin (PECH) is colorless and biocompatible. It has a low density and easy chemical modification through the Cl side group for polymer modification. PECH has commercial applications (fuel hoses, seals, and air ducts) due to its unique properties, including low-temperature flexibility and heat, oil, and ozone resistance. PECH derivatives are used as polymer electrolytes, gas separation membranes, and ion exchange membranes.[1] PECH can be polymerized by cationic ring-opening polymerization.[2] Since the foundation of "click" chemistry was laid,[3] there has been a surge in research describing the works of this precision chemical approach.[4–17]

In this work, PEG400-dip and PEG1450-diP were used as cross-linking bridges to chemically cross-link the PECH polymer. We aimed to achieve controlled and systematic cross-linking of PECH with PEG agents via "click" chemistry to examine the swelling behavior of the polymers depending on the organic solvent polarity and temperature to get practical data and clues about cross-linking efficiency. The obtained polymers with six different degrees of cross-linking were examined in four solvents (tetrahydrofuran [THF], DMF, methanol, and acetone) at 25 and 37°C.

1.2 EXPERIMENTAL METHODS AND MATERIALS

1.2.1 MATERIALS

PECH (Sigma–Aldrich/Mw ~700,000), sodium azide (Riedel-de Haen Agseelze-Hannover/99%), *N,N*-dimethylformamide (DMF) (Sigma-Aldrich/99%), polyethylene glycol 400 (Sigma–Aldrich/Mw = 380–420 g/mol), polyethylene glycol 1450 (Sigma–Aldrich/average mol wt 1.450), sodium hydride (Fluka/55–65% gas–volumetric), propargyl bromide (Sigma–Aldrich/80% in toluene), copper I bromide (ZAGxIndustry Chemicals/Mw 143.45 g/mol), methanol (Lab-Scan/99.9%), acetone (Tekkim/99.5%), and THF (Carlo Erba/99.9%).

1.2.2 INSTRUMENTS

Analytical balance (KERN ABJ 220-4NM/0.1 mg), magnetic hotplate-stirrer (DRAGONLAB ms7-h550-s), oven (Nüve FN300), vacuum oven (Nüve EV018), vacuum control box (iLMVAC VCZ 521), Fourier transform infrared spectroscopy (FT-IR) attenuated total reflectance (ATR) spectrometer (Shimadzu IRaffinity 1600–4000 cm^{-1}), elemental analysis (Elementar Vario Micro Cube), and TGA (TA, SDTQ600, 10°C/min, 25–500°C).

1.2.3 PARTIAL AZIDATION OF POLYEPICHLOROHYDRIN (PECH)

Partial azidation of PECH was done as given in Ref. [18]. Commercial PECH was dissolved in DMF and divided into three equal parts. In the first part, enough sodium azide was added to replace all the suspended chlorine atoms. In the second part, enough sodium azide was added to replace 50% of the chlorines. In the third part, enough sodium azide was added to replace 25% of the chlorines.

1.2.4 PROPARGYLATION OF POLYETHYLENE GLYCOLS (PEGS)

A similar synthesis is available in the literature.[19] 5 g of PEG oligomer samples (PEG400 and PEG1450) were added to NaH. Two end groups of the PEGs were made active. Thus, it was thought that the next reaction would be faster and more stable. For the propargylation of both ends of the PEGs, the amount of propargyl bromide to be added was calculated using the following calculation.

The amount of propargyl bromide required to prepare PEG400-diP:

$$5\,\mathrm{g} \times \frac{1\ \text{mol PEG}}{400\ \text{g PEG}} \times \frac{2\ \text{mol OH}}{1\ \text{mol PEG}} \times \frac{1\ \text{mol PrbBr}}{1\ \text{mol OH}} \times \frac{118.96\ \text{PrbBr}}{1\ \text{mol PrbBr}} \times \frac{100\ \text{g çözelti}}{80\ \text{g PrbBr}} \times \frac{1\ \text{mL}}{1.335} = 2.785\ \text{mL} \tag{1.1}$$

The amount of propargyl bromide required to prepare PEG1450-diP:

$$5\ \mathrm{g} \times \frac{1\ \text{mol PEG}}{1450\ \text{g PEG}} \times \frac{2\ \text{mol OH}}{1\ \text{mol PEG}} \times \frac{1\ \text{mol PrbBr}}{1\ \text{mol OH}} \times \frac{118.96\ \text{PrbBr}}{1\ \text{mol PrbBr}} \times \frac{100\ \text{g çözelti}}{80\ \text{g PrbBr}} \times \frac{1\ \text{mL}}{1.335} = 0.768\ \text{mL} \tag{1.2}$$

1.2.5 CROSS-LINKING WITH "CLICK" CHEMISTRY

Cross-linked PECHs (PECH25-PEG400, PECH50-PEG400, PECH100-PEG400, PECH25-PEG1450, PECH50-PEG1450, and PECH100-PEG 1450) were obtained by the "click" chemistry reaction of PECH azides (PECH25, PECH50, and PECH100) and PEG dipropargyls (PEG400-diP and PEG1450-diP). For this purpose, propargylated PEGs were used excessively for each modified PECH according to the azidation rate. Thus, the goal was to create maximum cross-linking. It was aimed to increase the cross-linking rate by extending the reaction time and increasing the temperature to 100°C. The gel product was added to the dry THF to extract any free PEG and PECH molecules. Then, the product was transferred into a vacuum oven and dried at 60°C for 2 days. The yield was determined gravimetrically.

The amount of PEG400-diP required to prepare PECH25-PEG400:

$$3\text{ g PECH}\times\frac{8\text{ g PECH}}{100\text{ solution}}\times\frac{1\text{ mol PECH}}{92.5\text{ g PECH unit}}\times\frac{14\text{ mol azide}}{100\text{ mL}}\times\frac{1\text{ mol PEGdip}}{2\text{ mol azide}}\times\frac{450\text{ PEGdip}}{1\text{ mol PEGdip}}=0.0817\,g\ (1.3) \quad (1.3)$$

The amount of PEG400-diP required to prepare PECH50-PEG400:

$$3\text{ g PECH}\times\frac{8\text{ g PECH}}{100\text{ g solution}}\times\frac{1\text{ mol PECH}}{92.5\text{ g PECH birim}}\times\frac{22\text{ mol azide}}{100\text{ mL}}\times\frac{1\text{ mol PEGdip}}{2\text{ mol azide}}\times\frac{450\text{ PEGdip}}{1\text{ mol PEGdip}}=0.1284\text{ g} \quad (1.4)$$

The amount of PEG400-diP required to prepare PECH100-PEG400:

$$3\text{ g PECH}\times\frac{8\text{ g PECH}}{100\text{g solution}}\times\frac{1\text{mol PECH}}{92.5\text{g PECH unit}}\times\frac{22.26\text{ mol azide}}{100\text{ mL}}\times\frac{1\text{ mol PEGdip}}{2\text{ mol azide}}\times\frac{450\text{ PEGdip}}{1\text{ mol PEGdip}}=0.1299\text{ g} \quad (1.5)$$

The amount of PEG400-diP required to prepare PECH25-PEG1400:

$$1.5\text{g PECH}\times\frac{8\text{ g PECH}}{100\text{ g solution}}\times\frac{1\text{ mol PECH}}{92.5\text{ g PECH unit}}\times\frac{14\text{ mol azide}}{100\text{ mL}}\times\frac{1\text{ mol PEGdip}}{2\text{ mol azide}}\times\frac{1500\text{ PEGdip}}{1\text{ mol PEGdip}}=0.136\text{ g} \quad (1.6)$$

The amount of PEG400-diP required to prepare PECH50-PEG1400:

$$1.5\text{ g PECH}\times\frac{8\text{ g PECH}}{100\text{ g solution}}\times\frac{1\text{ mol PECH}}{92.5\text{ g PECH unit}}\times\frac{22\text{ mol azide}}{100\text{ mL}}\times\frac{1\text{ mol PEGdip}}{2\text{ mol azide}}\times\frac{1500\text{ PEGdip}}{1\text{ mol PEGdip}}=0.214\text{ g} \quad (1.7)$$

The amount of PEG400-diP required to prepare PECH100-PEG1400:

$$1.5\text{g PECH}\times\frac{8\text{ g PECH}}{100\text{ solution}}\times\frac{1\text{mol PECH}}{92.5\text{ g PECH unit}}\times\frac{22.26\text{ mol azide}}{100\text{ mL}}\times\frac{1\text{ mol PEGdip}}{2\text{ mol azide}}\times\frac{1500\text{ PEGdip}}{1\text{ mol PEGdip}}=0.216\text{ g} \quad (1.8)$$

1.2.6 SWELLING RATIO OF THE CROSS-LINKED POLYMERS

The swelling behaviors of the cross-linked products in THF, DMF, methanol, and acetone were examined at 25 and 37°C. For this purpose, the dry-weighed polymers were left in containers containing the solvents. After a predetermined time (1, 2, 4, 8, 12, 24, and 48 h, etc.), they were removed from the solvent, dried, and weighed. The degrees of swelling were calculated using the equation below. This process continued for four different solvents and two different temperatures for six different cross-linked polymers until swelling ceased.

$$Swelling(\%) = \frac{W - Wo}{Wo} \times 100 \tag{1.9}$$

W: Swelling polymer weight

Wo: Dried polymer weight

1.3 RESULTS AND DISCUSSION

1.3.1 PARTIAL AZIDATION OF PECH

With the azidation reaction shown in Scheme 1.1, PECH polymers with different azidition ratios were obtained. FT-IR analysis was performed to detect the azidation. For this purpose, the precipitated polymer was washed with water, and the unreacted azide was completely removed and

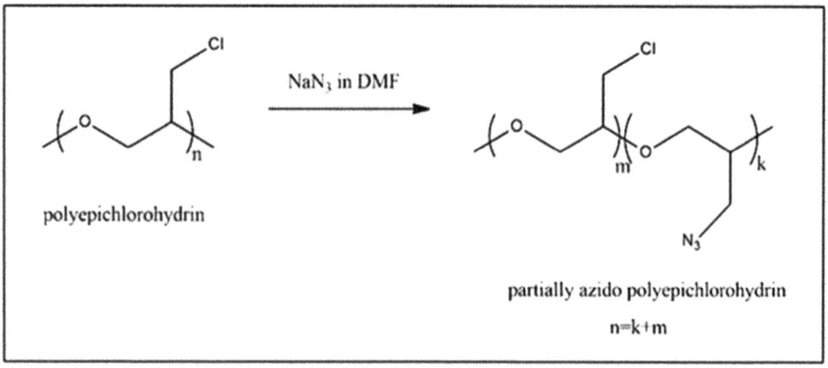

SCHEME 1.1 Reaction outline of the partial azidation of PECH.

then dried in an oven at 60°C for 2 days. Figure 1.1 shows the FT-IR spectrum of the PECH azide. The azide vibration band was shown at 2098 cm^{-1}. The azide vibration band observed at 2098 cm^{-1} shows that the azidation reaction had been successfully achieved.[20] Figure 1.1 shows that the intensity of the C–Cl band at 748 cm^{-1} decreases after azidation. Elemental analysis was performed to determine the azidation rate. With this analysis, the nitrogen content in the polymer samples was determined and the azidation rate was calculated stoichiometrically as azidation %. The results of the elemental analysis are given in Table 1.1.

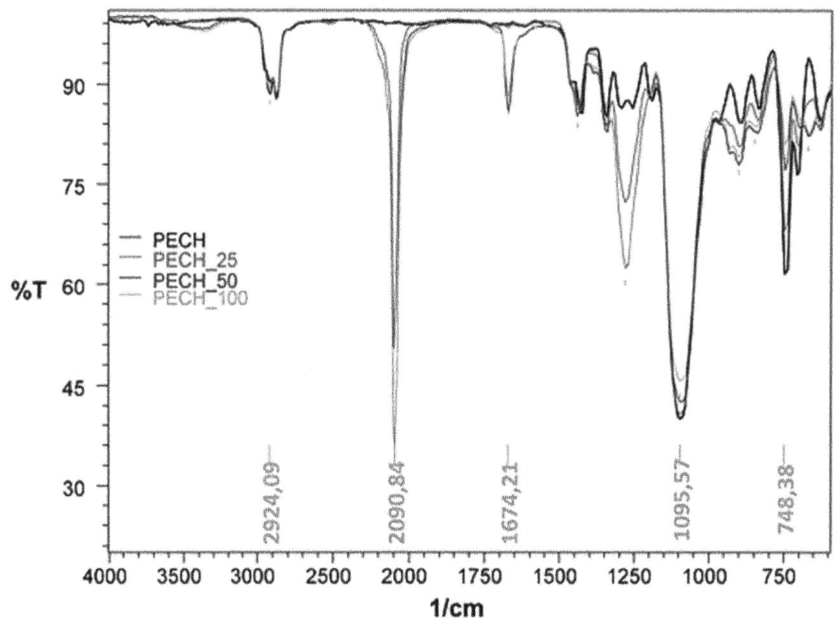

FIGURE 1.1 FT-IR spectra of PECH and modified PECHs.

TABLE 1.1 PECH Azidation Percentage.

PECH-N_3	N%	Azidation%
25% (intent)	10.60	24.99
50% (intent)	13.89	32.74
100% (intent)	13.65	32.17
%100 N_3 (theoretical)	42.42	

1.3.2 PROPARGYLATION OF PEGS

The reaction pathway of propargylation of PEG was shown in Scheme 1.2. As shown in the FT-IR spectrum, the disappearance of the broad O–H tensile band at approximately 3449 cm^{-1} after treatment with NaH was attributed to the conversion of hydroxy groups at the chain ends to alkoxy. The weak bands observed at approximately 1967 and 3248 cm^{-1} belong to the bond between the C≡C triple bond and the hydrogen attached to the carbon-carbon triple bond (C≡C–H), respectively (Figure 1.2).

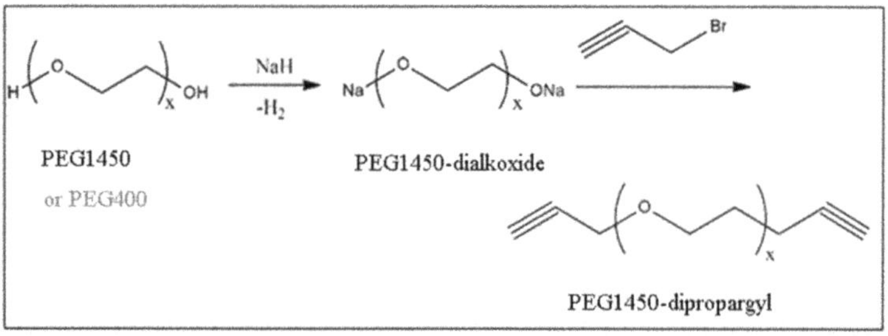

SCHEME 1.2 Reaction pathway of the dipropargylation of PEG.

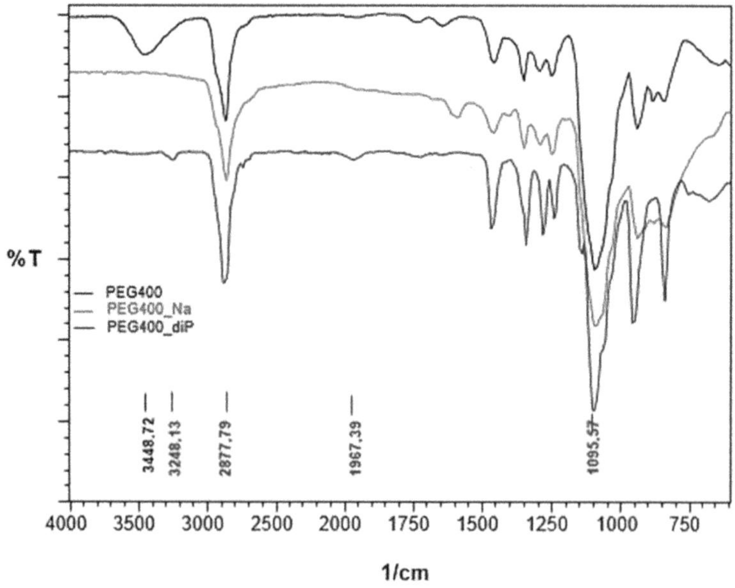

FIGURE 1.2 FT-IR spectra of PEG, PEG400-Na, and PEG400-diP.

1.3.3 SYNTHESIS OF CROSS-LINKED PECH

The reaction outline of the synthesis of cross-linked PECH was shown in Schemes 1.3 and 1.4. A few hours after the reaction started, it was observed that the viscosity started to increase, and the solutions were completely gelled at the end of the second day. In the FT-IR spectra (Figure 1.3), the peaks at about 3100 cm^{-1} indicated the -CH proton of the triazole ring.[21] This very specific band is an indisputable indication that the click reaction is achieved. Decomposition temperatures (Td) of PECH cross-linked polymers are shown in Table 1.2. Also, Figure 1.4 indicated TGA thermograms of PECH cross-linked polymers. The Td value for PECH based on TGA analysis in the literature is 364°C.[22] Similarly, it is given as 383°C in a single step for the Td of PEG.[23] It could be concluded that the first of the two-step decay in TGA diagrams to PECH and the second to PEG units. As shown in Figure 1.4, the mass of the PEG1450 bridges used in the PECH50-PEG1450 and PECH100-PEG1450 polymers is greater by mass than the PECH polymer forming the main chains. Looking at the peak heights, it could be said that there was a linear relationship between the cross-link density and the PEG content of the obtained cross-linked polymer. In other words, it was seen that the PEG unit was the least in PECH25-PEG400 and the most in PECH100-PEG400. However, PECH main chains remained massively dominant in all three of the cross-linked products obtained with PEG400 bridges. For PECH-PEG400 samples, the values of swelling degrees decrease with cross-link density in all the solvents at both 25 and 37°C (Figure 1.5). For the PECH-PEG400 samples, regardless of cross-linking degrees, the values of swelling degree increase with temperature increase in all the solvents (Figure 1.6). For the samples with the same cross-link density, the values of swelling degrees increase with the length of bridges in DMF and at both temperatures (Figure 1.7). Figure 1.8 showed the effect of the length of the bridge on the swelling for THF. The effect of the length of the bridge on the swelling of acetone is shown in Figure 1.9. Figure 1.10 indicates the Effect of the length of the bridge on the swelling of methanol. The swelling degree of all samples in DMF at both temperatures was higher than those in other solvents (Figure 1.11). Generally, trends in swelling degree were DMF > THF > MeOH > Acetone (Figure 1.11).

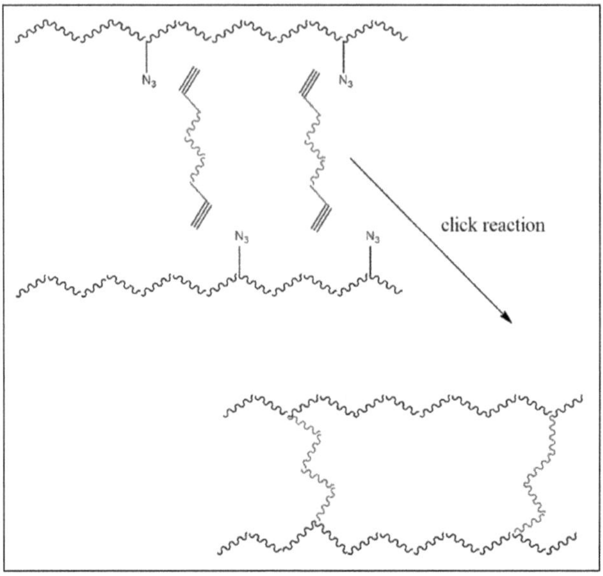

SCHEME 1.3 Symbolic representation of cross-linked PECH.

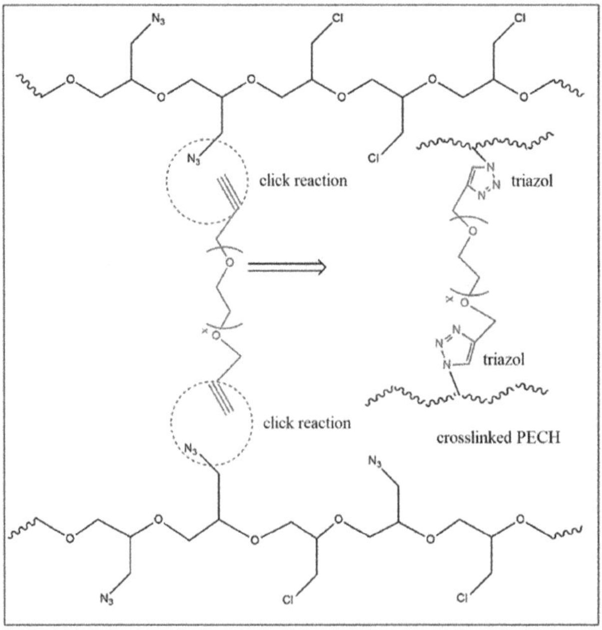

SCHEME 1.4 The cross-linking reaction by "click" reaction.

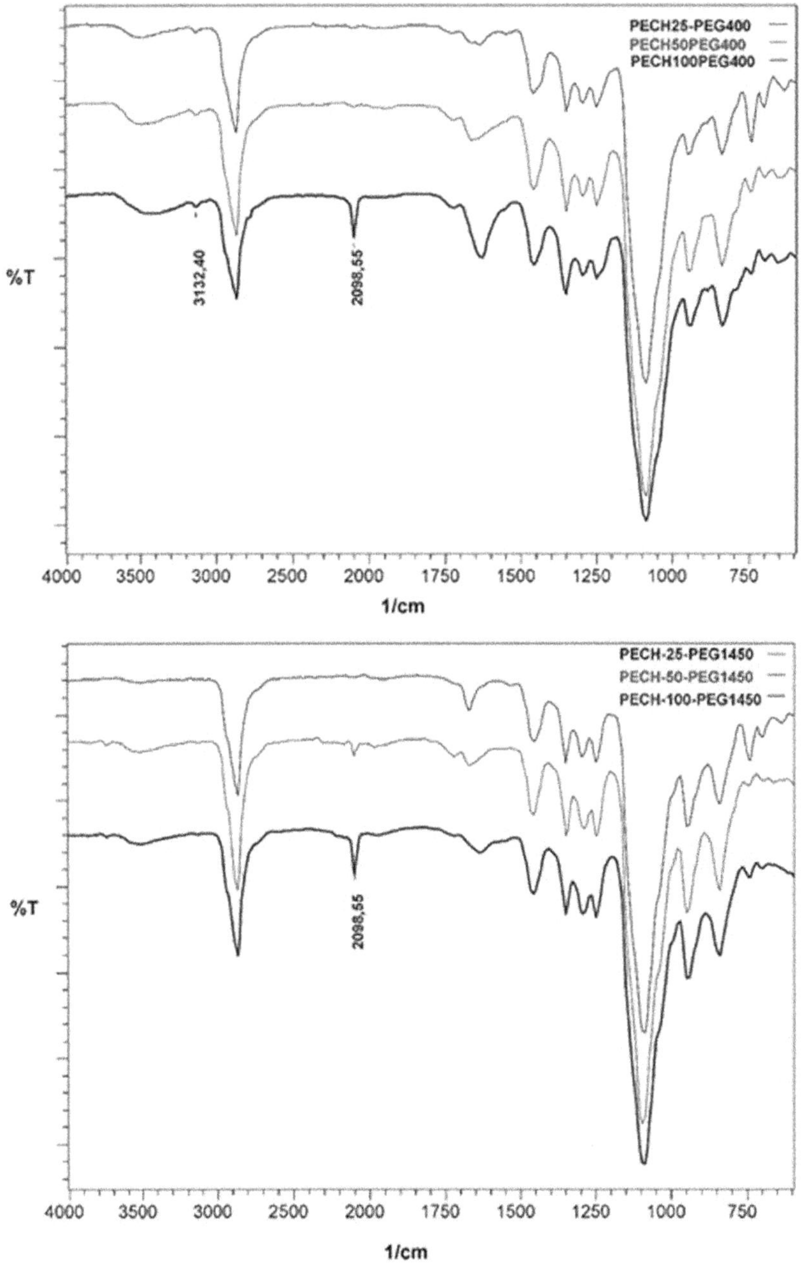

FIGURE 1.3 FT-IR spectra of PECH cross-linked polymers.

TABLE 1.2 Decomposition Temperatures (Td) of PECH Cross-linked Polymers.

Code	Td1 (oC)	Td2 (oC)
PECH25-PEG400	278.87	348.01
PECH50-PEG400	291.92	357.50
PECH100-PEG400	321.09	353.18
PECH25-PEG1450	283.83	400.21
PECH50-PEG1450	304.01	406.06
PECH100-PEG1450	360.09	410.13

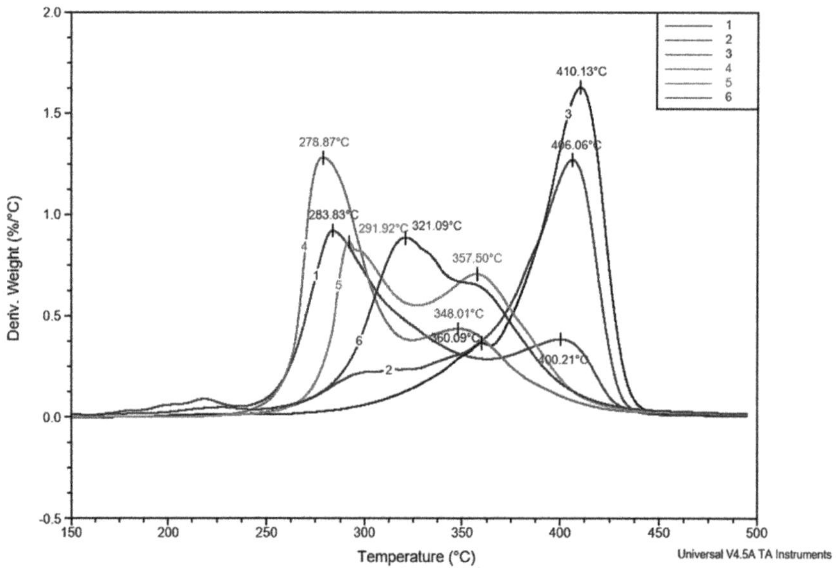

FIGURE 1.4 TGA thermograms of PECH cross-linked polymers (1. PECH25- PEG1450, 2. PECH50-PEG1450, 3. PECH100-PEG1450, 4. PECH25-PEG400, 5. PECH50-PEG400, 6. PECH100-PEG400).

1.4 CONCLUSION

PECH polymers cross-linked with two types of PEG bridges (PECH25-400, PECH50-400, PECH100-400, PECH25-1450- PEG50-1450, PECH100-1450) were obtained successfully through "click" chemistry. As the percentage of cross-link increases, the swelling ratio decreases in all samples inversely. As the temperature increased, the swelling percentages increased in all samples. As the bridge length increased, the swelling percentage

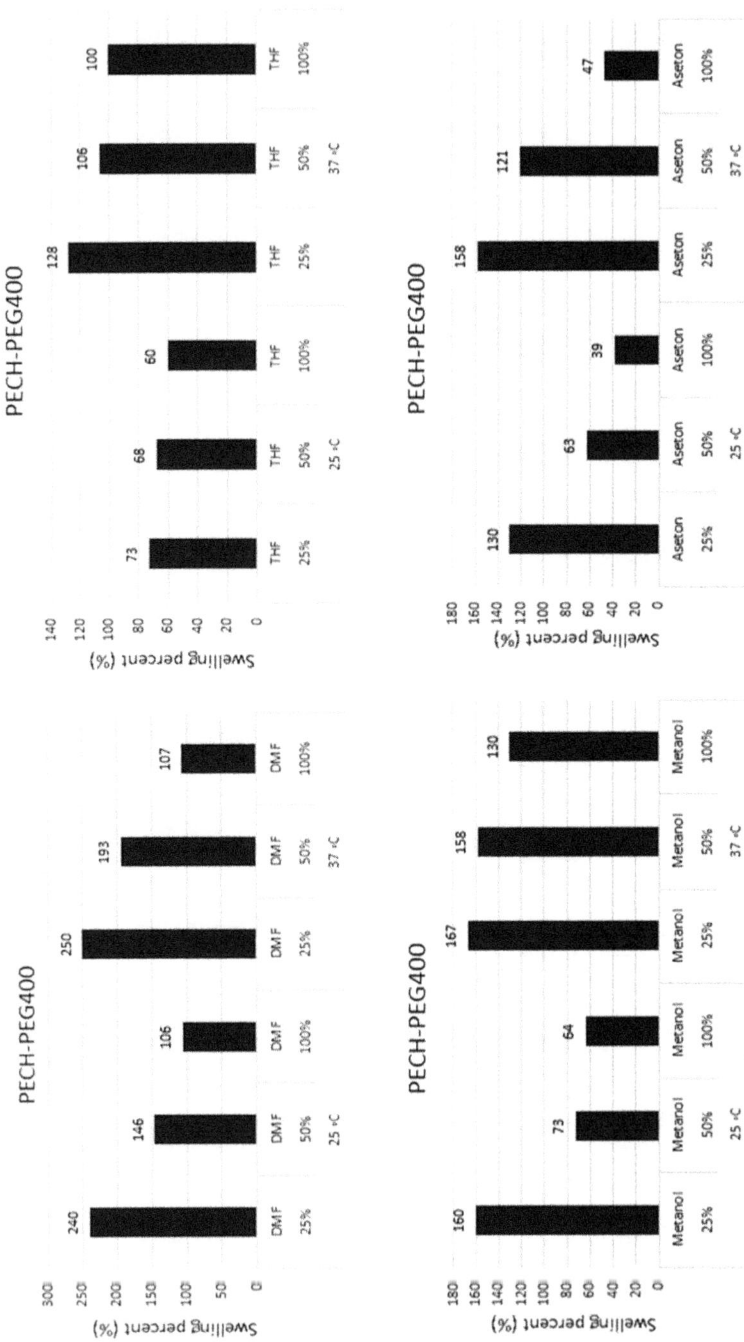

FIGURE 1.5 Effect of the cross-link density on the swelling.

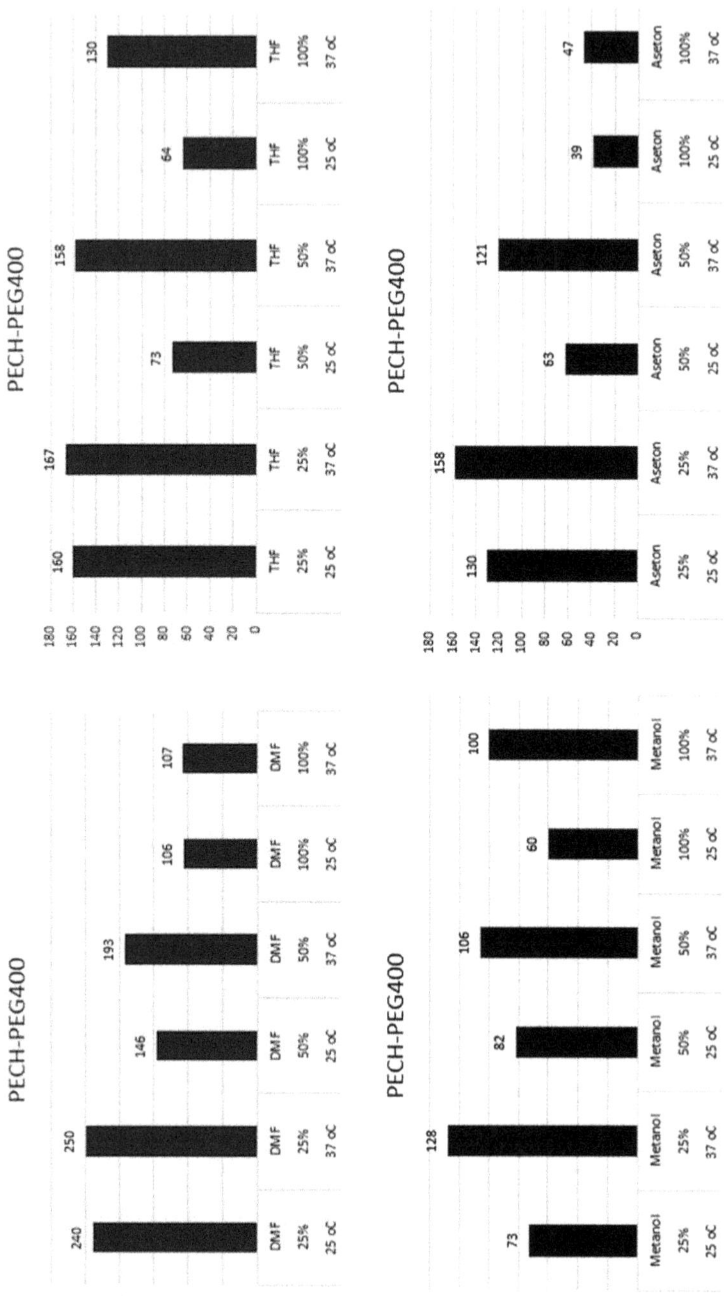

FIGURE 1.6 Effect of the temperature on the swelling.

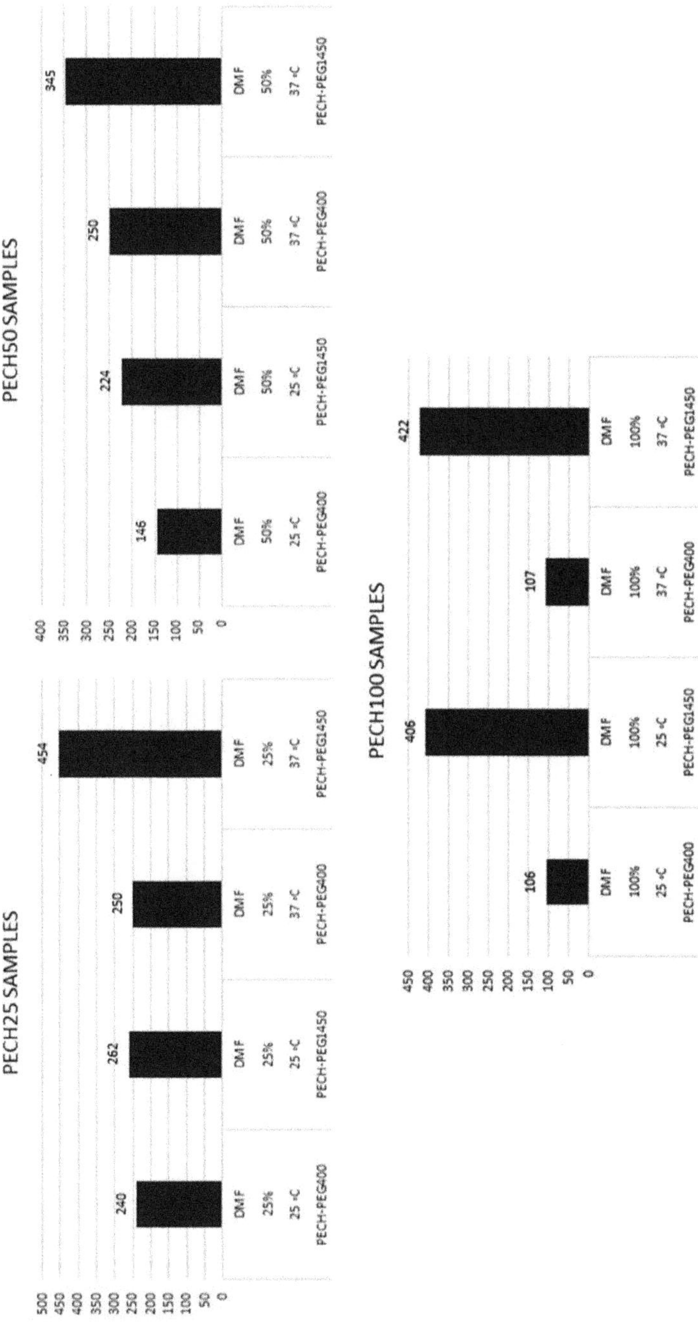

FIGURE 1.7 Effect of length of the bridge on the swelling for DMF.

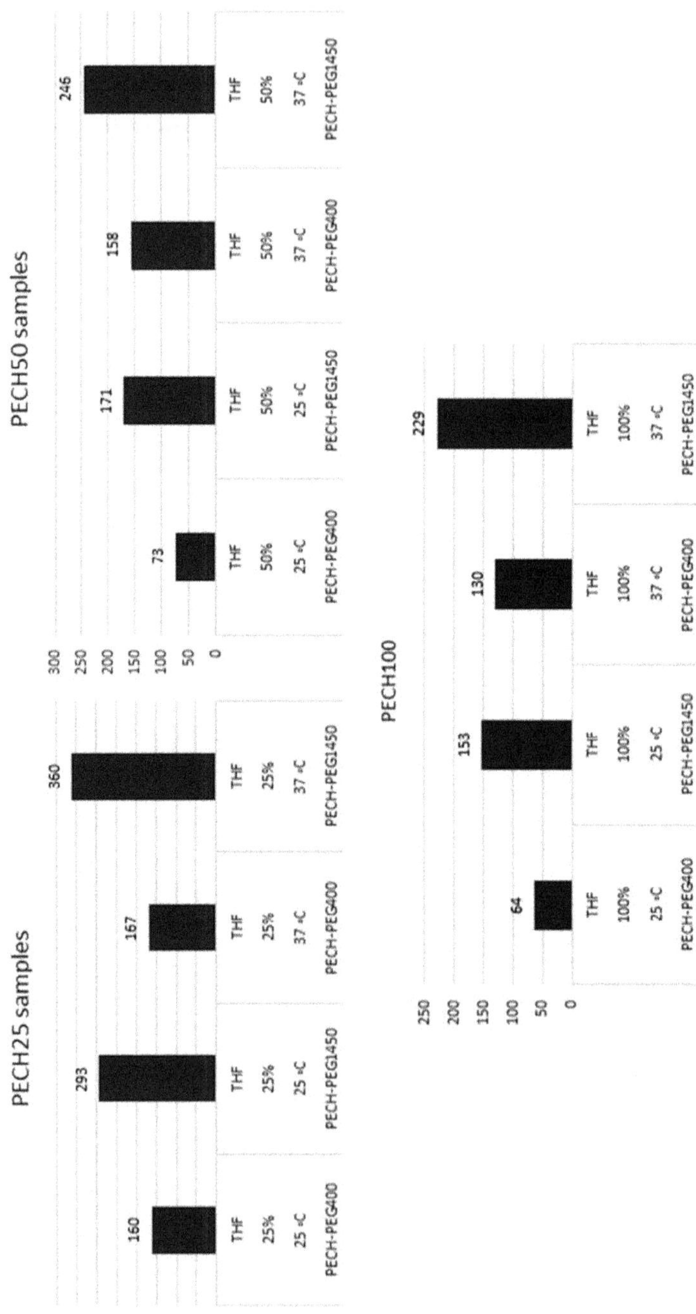

FIGURE 1.8 Effect of length of the bridge on the swelling for THF.

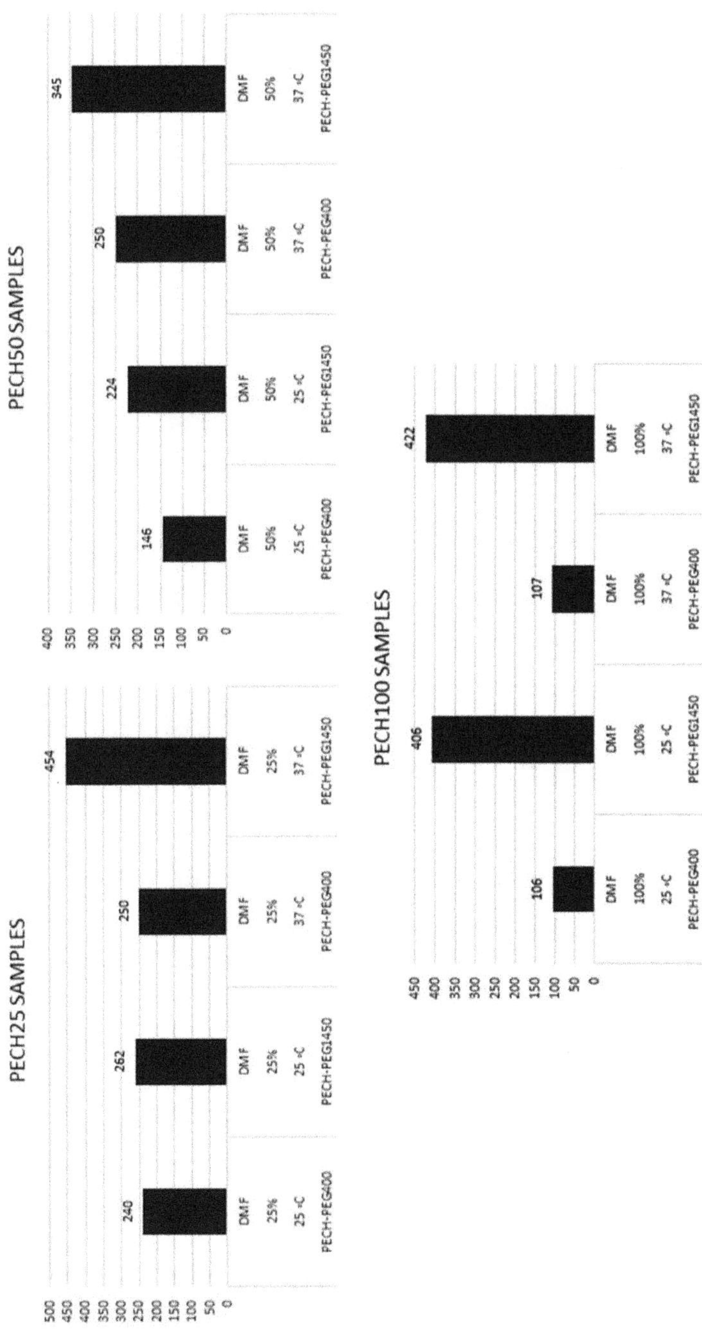

FIGURE 1.7 Effect of length of the bridge on the swelling for DMF.

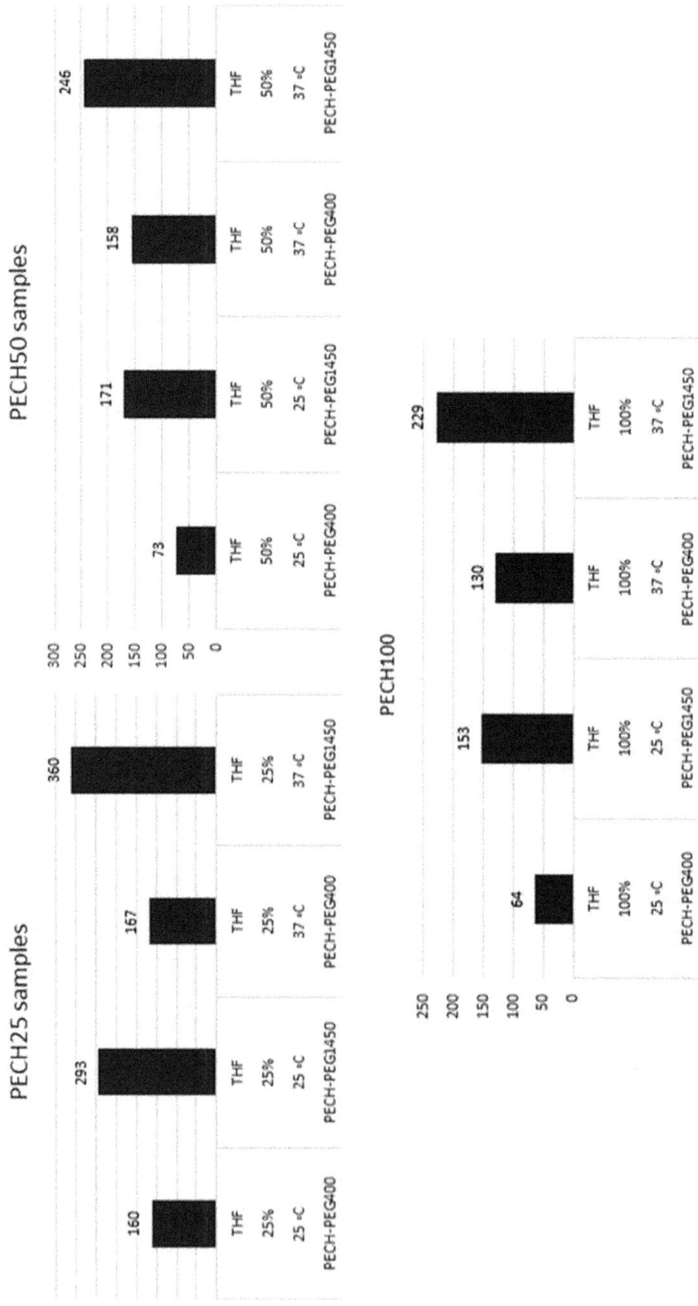

FIGURE 1.8 Effect of length of the bridge on the swelling for THF.

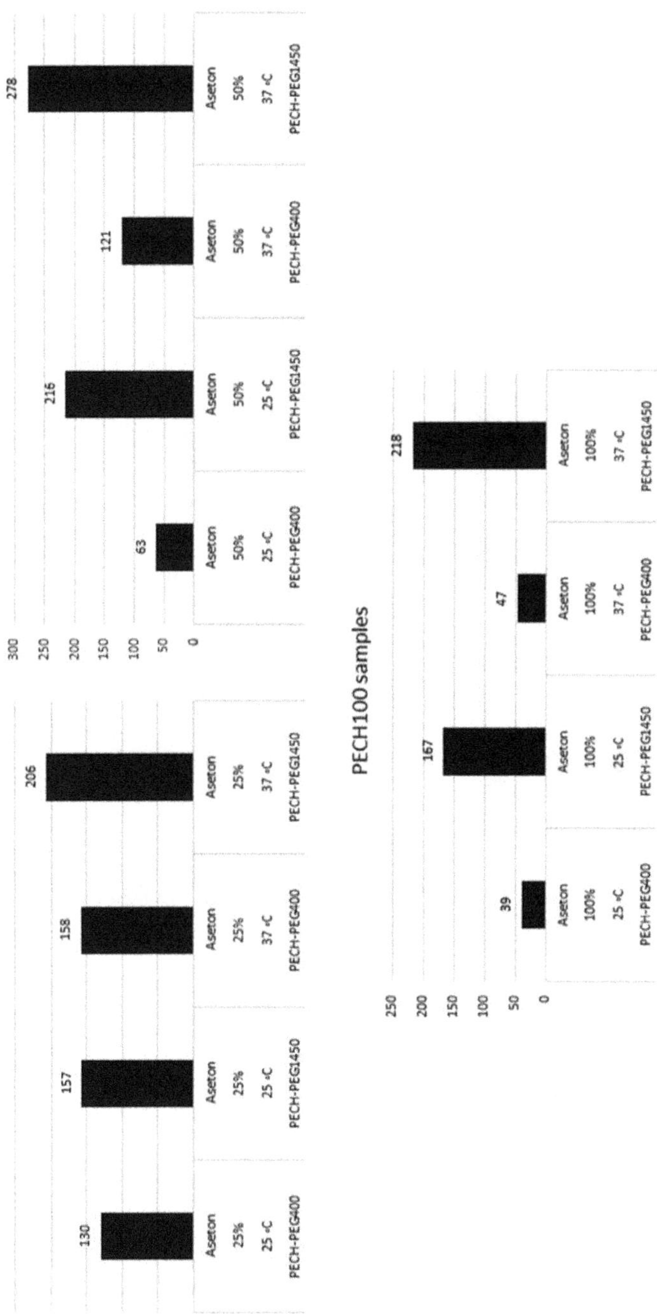

FIGURE 1.9 Effect of length of the bridge on the swelling for acetone.

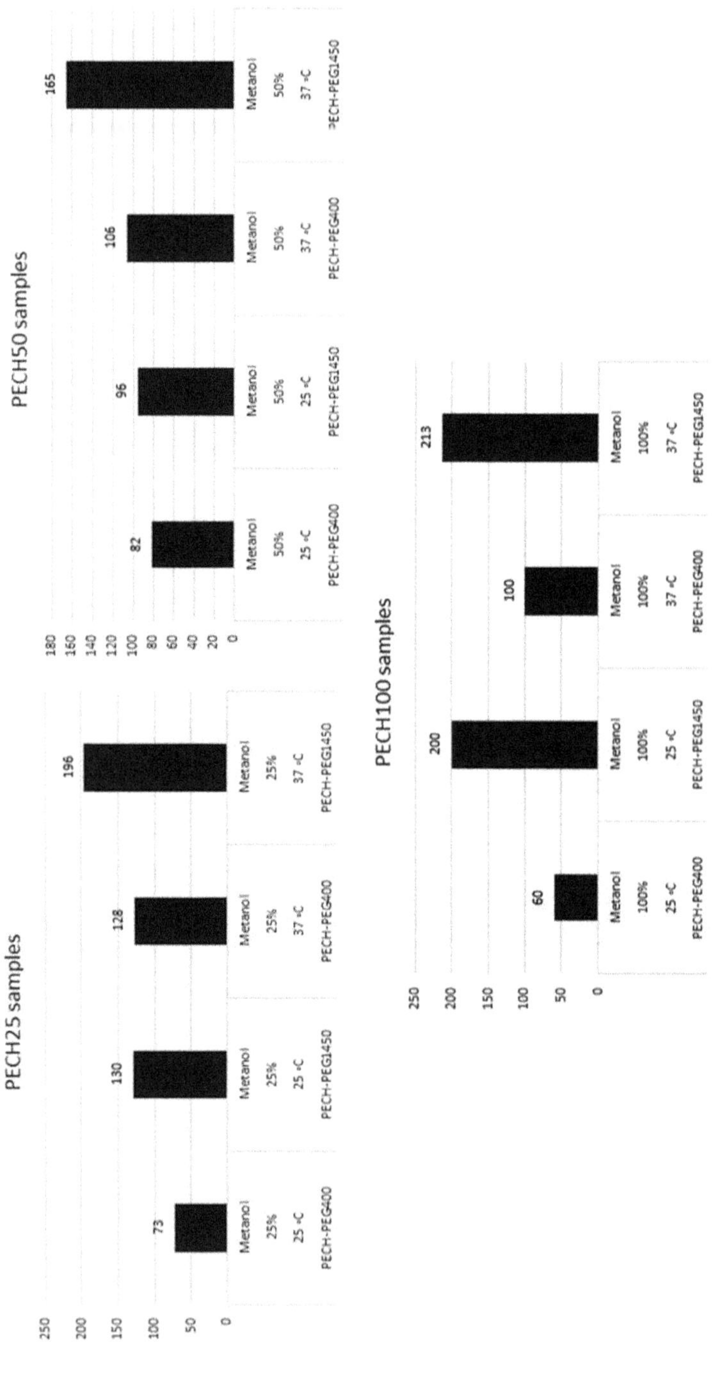

FIGURE 1.10 Effect of length of the bridge on the swelling for methanol.

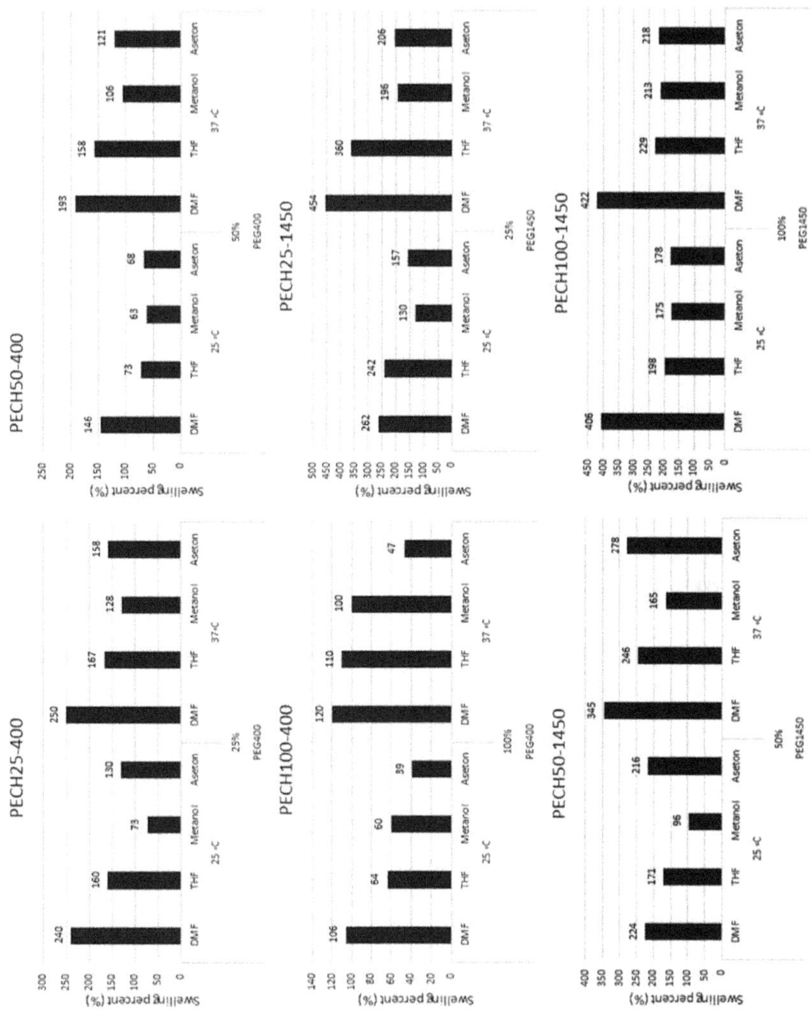

FIGURE 1.11 The effect of the type of solvent on the swelling.

increased in all samples. The degrees of swelling in DMF were highest, the general trend was like DMF>THF>methanol>acetone.

ACKNOWLEDGMENT

The research project was supported financially by the Scientific Research Projects Coordination Unit of Ordu University (Project Number: BAP 2208).

KEYWORDS

- **click reaction**
- **cross-linking**
- **polyepichlorohydrin**
- **polyethylene glycol**
- **swelling percentage**

REFERENCES

1. Shukla, G.; Ferrier, R. C. The Versatile, Functional Polyether, Polyepichlorohydrin: History, Synthesis, and Applications. *J. Polym. Sci* **2021,** *59,* 2704–2718.
2. Öztürk, T.; Atalar, M. N.; Göktaş, M., Hazer, B. One-Step Synthesis of Block-Graft Copolymers via Simultaneous Reversible-Addition Fragmentation Chain Transfer and Ring-Opening Polymerization Using a Novel Macroinitiator. *J. Polym. Sci. Part A: Polym. Chem.* **2013,** *51,* 2651–2659.
3. Kolb, H. C.; Finn, M. G.; Sharpless, K. B. Click Chemistry: Diverse Chemical Function from a Few Good Reactions. *Angewandte Chemie Int. Ed.* **2001,** *40,* 1983–2181.
4. Öztürk, T.; Türkoğlu, H. Synthesis and Characterization of the Graft Copolymer Including Polyβ-butyrolactone and Polyvinyl Chloride by Ring-opening Polymerization and "click " Chemistry. *J. Macromole. Sci., Part A: Pure Appl. Chem.* **2022,** *59,* 871–878.
5. Savaş, B.; Meyvacı, E.; Öztürk, T.; Ağırtaş, M. S. Synthesis and Characterization of Poly(ε-caprolactone) Tetra-arm Star Polymer using Tetra Terminal Alkynyl-substituted Phthalocyanine by the Combination of Ring-opening Polymerization and "Click" Chemistry. *Ovidius Univ. Annal. Chem.* **2022,** *33,* 17–22.
6. Savaş, B.; Öztürk, T. Synthesis and Characterization of Poly(vinyl chloride-g-methyl methacrylate) Graft Copolymer by Redox Polymerization and Cu Catalyzed

Azide-alkyne Cycloaddition Reaction. *J. Macromole. Sci., Part A: Pure Appl. Chem.* **2020,** *57,* 819–825.

7. Binder, W. H.; Sachsenhofer, R. 'Click' Chemistry in Polymer and Materials Science. *Macromole. Rapid Commun.* **2007,** *28,* 15–54.
8. Savaş, B.; Çatıker, E.; Öztürk, T.; Meyvacı, E. Synthesis and Characterization of Poly(α-methyl β-alanine)-poly(ε-caprolactone) Tri Arm Star Polymer by Hydrogen Transfer Polymerization, Ring-opening Polymerization and "click" Chemistry. *J. Polym. Res.* **2021,** *28,* 30.
9. Öztürk, T.; Cavicchi, C. A. Synthesis and Characterization of Poly(epichlorohydrin-g-ε-caprolactone) Graft Copolymers by "Click" Chemistry. *J. Polym. Mater.* **2018,** *35,* 209–220.
10. Moses, J. E.; Moorhouse, A. D. The Growing Applications of Click Chemistry. *Chem. Soc. Rev.* **2007,** *36,* 1249–1262.
11. Öztürk, T.; Kılıçlıoğlu, A.; Savaş, B.; Hazer, B. Synthesis and Characterization of Poly(ε-caprolactone-co-ethylene glycol) Star-type Amphiphilic Copolymers by "click " Chemistry and Ring-opening Polymerization. *J. Macromole. Sci. Part A: Pure Appl. Chem.* **2018,** *55,* 588–594.
12. Çatıker, E.; Meyvacı, E.; Atakay, M.; Salih, B.; Öztürk, T. Synthesis and Characterization of Amphiphilic Triblock Copolymers Including β-alanine/α-methyl-β-alanine and Ethylene glycol by "click" Chemistry. *Polym. Bullet.* **2019,** *76,* 2113–2128.
13. Öztürk, T.; Meyvacı, E.; Bektaş, H.; Menteşe, E. Synthesis and Characterization of Ring-type and Branched Polymers Including Polyethylene Glycols by "click" Chemistry. *SN Appl. Sci.* **2019,** *1,* 343.
14. Hein, C. D.; Liu, X. M.; Wang, D. Click Chemistry, a Powerful Tool for Pharmaceutical Sciences. *Pharm. Res.* **2008,** *25,* 2216–2230.
15. Öztürk, T.; Meyvacı, E. Synthesis and Characterization Poly(Ɛ-Caprolactone-b-Ethylene Glycol-b-Ɛ-Caprolactone) ABA Type Block Copolymers via "Click" Chemistry and Ring-Opening Polymerization. *J. Macromole. Sci., Part A: Pure Appl Chem.* **2017,** *54,* 575–581.
16. Tunca, U. Click and Multicomponent Reactions Work Together for Polymer Chemistry. *Macromole. Chem. Phy.* **2018,** *219,* 1800163.
17. Barner-Kowollik, C.; Du Prez, F. E.; Espeel, P.; Hawker, C. J.; Junkers, T.; Schlaad, H.; Camp, W. V. "Clicking " Polymers or Just Efficient Linking: What Is the Difference? *Angewandte Chemie Int. Ed.* **2011,** *50,* 60–62.
18. Öztürk, T.; Cavicchi, C. A. Synthesis and Characterization of Poly(epichlorohydrin-g-ε-caprolactone) Graft Copolymers by "Click" Chemistry. *J. Polym. Mater.* **2018,** *35,* 209–220.
19. Asan, N.; Öztürk, T.; Kudu, H.; Meyvacı, E.; Çatıker, E. Cross-linking of PVC by Various Lengths of PEG via "Click" Chemistry. *Karaelmas Fen ve Mühendislik Dergisi / Karaelmas Sci. Eng. J.* **2022,** *12,* 59–66.
20. Yu-Han C.; Guohua, C.; Duu-Jong, L. Synthesis of Low Surface Energy Thin Film of Polyepichlorohydrin-triazole-ols. *J. Coll. Inter. Sci.* **2020,** *575,* 452–463.
21. Moses, E. J.; Moorhouse, A. D. The Growing Application of Click Chemistry. *Chem. Soc. Rev.* **2007,** *36,* 1249–1262.

22. Nelson, L.; Souza, G. D.; Munk, M.; Humberto, M.; Luiz Fernando, C. Synthesis and Properties of (Alkylthio)methyl-Substituted Poly(oxyalkylene)s and (Alkylsulfonyl) methyl-Substituted Poly(oxyalkylene)s. *Polym. Bullet.* **2018,** *75,* 4627–4639.
23. Tingting, Q.; Li, J.; Feng, W.; Nian, H. Hybrid Magnetic Nanovectors Promote Selective Glioblastoma Cell Death through a Combined Effect of Lysosomal Membrane Permeabilization and Chemotherapy. *Scientif. Rep.* **2017,** *7,* 44710.

CHAPTER 2

Synthesis of the Tl-Based Superconducting Materials by the Sol–Gel Method Iodine Vapor Pressure

METSKHVARISHVILI IOSEB[1,2], LOBZHANIDZE TEA[3], SURMANIDZE DAVITI[1,3], DGEBUADZE GURAM[1], BENDELIANI BEZHAN[1], METSKHVARISHVILI MAGDA[4], and GABUNIA VAKHTANG[5]

[1]*Ilia Vekua Sukhumi Institute of Physics and Technology, Laboratory of Cryogenic Technique and Technologies, Mindeli, Tbilisi, Georgia*

[2]*Georgian Technical University, Faculty of Informatics and Control Systems, Department of Microprocessor and Measurement Systems, Kostava, Tbilisi, Georgia*

[3]*Ivane Javakhishvili Tbilisi State University, Faculty of Exact and Natural Sciences, Department of Chemistry, Chavchavadze, Tbilisi, Georgia*

[4]*"Talga" Institute of Georgian Technical University, Zurab Anjaparidze, Tbilisi, Georgia*

[5]*Petre Melikishvili Institute of Physical and Organic Chemistry of the Ivane Javakhishvili Tbilisi State University, Jikia, Tbilisi, Georgia*

ABSTRACT

It is known that the formation of the Tl-based superconducting materials critically relies on the precursor employed and synthesis conditions. The substitution or addition of various dopants in Tl-1223 phases has many fascinating queens; besides phase stability, the physical properties of the

Advances in Engineering Materials: Preparation for Sustainable Process Development. Tamara Tatrishvili and A. K. Haghi (Eds.)

Tl-1223 superconductor also improve upon chemical substitution. In the present work, we used the *in situ* polymerization method to synthesize high-purity and reaction-ability precursors. Prior reports have yet to detail such an approach for Tl-based superconductors. Also, the novelty of the research is that as dopants, we use iodine pentoxide (I_2O_5), whose transition temperature (T_c) to steam is lower than Tl_2O_3 oxide. Using this approach, we controlled the vapor pressure of Tl_2O_3 in a sealed quartz tube using I_2O_5. As far as we know, such methods in synthesizing Tl-based superconductor materials and their influence on phase formation, transport, and magnetic properties were investigated first. In summary, our results showed that I_2O_5 positively enhances the phase purity and superconductivity properties of the Tl-1223 phases.

2.1 INTRODUCTION

Tl-1223 stands out among the members of the Tl-based series as a preferred candidate for superconductors in both fundamental and applied research. This is due to the high temperature (T_c), critical current density (J_c), and low microwave surface resistance R_s exhibited by Tl-based superconductors. Consequently, they hold promise for applications such as radio-frequency radiation, Josephson-effect-based devices, and large quasi-permanent magnet devices.[1–5]

The high-temperature superconductors Tl-1223 show a superconducting transition temperature (T_c) above 120 K,[6,7] and when synthesized at high pressure, 133.5 K can be reached as the critical temperature.[8,9] The formula of the $TlBa_2Ca_{n-1}Cu_nO_{2n+2+\delta}$ system and its crystal structure are comparable to those of the $HgBa_2Ca_{n-1}Cu_nO2_{n+2+\delta}$ system. (n) is the number of Cu–O layers, and in both systems, it is essential and critically depends on the properties of the $Ba_2Ca_2Cu_3O_x$ precursor to accomplish a high-purity superconductivity phase. As is known, the solid-state reaction method, historically used in synthesizing precursors, cannot provide possibilities of stoichiometry control and makes it impossible to obtain a homogeneous superconductor with the preferred phase composition.[10,11] Because high-temperature superconducting materials have multiple constituent ions, Y-based superconductors have three ions, four or five for Bi-based, Tl-based, and Hg-based superconductors, and occasionally more than five. Based on this, the wet chemistry techniques are more exquisite than solid-state reaction methods. Solid-state reaction and co-precipitation techniques exacerbate issues in solid solutions. As a result, differences in compositional homogeneity and small phase purity can be seen in the properties of polycrystalline superconductors despite their same overall composition. The sol–gel process's advantages are forming bulk superconducting materials from gels and achieving homogeneous mixing of cations on an atomic scale.[11–14]

Formatting the Tl-1223 phase is challenging only under pressure created from Tl_2O_3 oxides at a temperature of boiling and decomposition. Thus, the nature of the grain boundaries can be enhanced, the formation of the superconducting phase can be encouraged, or the effective pinning centers can be introduced by substituting or adding metals or oxides to Tl-based superconductors. The influence of rare-earth elements on superconductivity properties is presented by Khan et al.[15] They demonstrated that doping with *Y* reduces the superconductivity parameters of ξ_c, J_c, and V_F, while doping with Er and Gd increases the values of these parameters. Y, Er, and Gd doping, however, suppresses the values of magnetic fields B(T) and critical current density, $J_c(0)$. Furthermore, the doping of rare-earth elements increases the values of the GL parameter κ and the magnetic field penetration depth. The effect of Li_2O doping on the microstructural, magnetic, and transport properties of Tl-2223 superconductors is discussed in the paper.[11] They discovered that although a 20% lithium concentration increases the critical current density, J_c, Li_2O slightly lowers the superconducting T_c. According to K. A. Jasim,[17] partial substitution of Cd for Tl in Tl-1223 results in a temperature increase from 113 to 123 K, as well as modifications to the mass density ρ_m, C/a, and volume fraction $V_{Ph(1223)}$. Substituted Tl by Ti showed increases in the volume fraction of the Tl-1223 phase from 85.6–88.2%, and critical temperature increased from 123 to 125.3 K where Ti ≈ 0.10 wt.%.[18] As discussed in Ref. [19], magnesium (Mg) doping promotes inter-plane coupling and corresponding enhancement of k_F, ξ_c, and V_F from these superconductivity parameters.[19] Numerous academic institutions have investigated the impact of nanoparticles on superconductivity properties.[20,21] $(MgO)_x$/CuTl-1223 nanoparticles–superconductor composites were synthesized and characterized in Ref. [20]. They found that the crystal structure of investigation samples did not change with the addition of MgO nanoparticles, but superconductivity was suppressed with the addition of nanoparticles. Besides, the influence of Nanosized ZnO on the physical properties[21] showed that nanozinc oxide addition does not affect the formation of the tetragonal structure of the CuTl-1223 phase and T_c critical temperature. On the contrary, the transport critical current density, J_c, of the synthesized samples depends on the nano-ZnO concentrations.

We report synthesizing the precursors via the sol–gel route in relation to this. We used the *in situ* polymerization method for synthesizing high-purity and reaction-ability precursors. Also, as dopants, we use iodine pentoxide (I_2O_5), whose T_c to steam is lower than Tl_2O_3 oxide. Using this approach, we can control the vapor pressure created by thallium trioxide in a sealed quartz tube and synthesize samples of higher pressures than we can make with thallium trioxide when it decomposes to 875°C degrees above. We must note that such an approach for Tl-based superconductors has not yet been reported.

The X-ray diffraction (XRD) results exhibited the patterns' crystalline structure. The infrared (Fourier Transform Infrared Spectroscopy [FT-IR]) spectrum investigated phonon vibration, and scanning electron microscopy (SEM) detected samples' surface morphology. The T_c and J_c were determined using the high harmonics method.[22–24] We use the high harmonics method for transport J_c measurements.[25–29]

The XRD results exhibited the patterns' crystalline structure. The infrared (FT-IR) spectrum investigated phonon vibration, and SEM detected samples' surface morphology. The T_c and critical current density (J_c) were determined using the high harmonics method.[22–24] We use the high harmonics method for transport critical current density, J_c, measurements.[25–29]

According to Bean's critical-state model, the amplitude of each harmonic is inversely proportional to the critical current density, J_c, $V_n \sim 1/J_c$, which was initially used to interpret the high harmonics.[25] It is common knowledge that if we apply sinusoidal excitation to a nonlinear system and observe how it responds to this excitation, the response signal will not be sinusoidal. Simultaneously, the response spectrum can be expressed as the sum of harmonic signals when $H = 0$; only odd harmonics are observed. Even harmonics also appear when a direct current (DC) field $H \neq 0$ is applied, and as the number of harmonics increases, the amplitudes of the high harmonics decrease. We use the third harmonic voltage amplitude of the high harmonic ($|\chi^3|$)[22–24] for investigation.

2.2 EXPERIMENTAL METHODS AND MATERIALS

We used a two-step method for the synthesis of $TlBa_2Ca_2Cu_3O_{8+\delta}$ samples. In the first step, Tl-free precursors were made. After this, Tl_2O_3

was added and mixed, the obtained powder was pressed into pellets, and the last sintering in the quartz tubes was done in the second stage. We must note that we used oxides and carbonate-containing oxides for the starting material ($BaCO_3$, $CaCO_3$, CuO, and Tl_2O_3).

In the previous work,[11] the precursor was synthesized using the modified sol–gel method with polyvinyl alcohol (PVA)/polyvinyl acetate (PVAc) as the complexing agent of metal cations in the gel process. This study uses the *in situ* polymerization method for synthesizing precursors, with ethylenediaminetetraacetic acid (EDTA) as the complexing agent in an aqueous solution.

$Ba_2Ca_2Cu_3O_x$ was synthesized by synthesizing $CaCO_3$, CuO, and $BaCO_3$ nitric, and acetic acids were used as solvents. The obtained solutions were then mixed in a stoichiometric ratio of 2:2:3. $BaCO_3$ was dissolved in acetic acid to prevent the precipitation of high concentrations of $BaCO_3$, EDTA was introduced in a 1:1 ratio, and the pH was adjusted to 3 using ammonia. Within the solution, copper cations form a complex. This particular compound showed the strongest complexation with the cations mentioned above. Subsequently, all solutions were combined with water, obtaining 10 g of the precursor phase within a total solution volume of approximately 200 mL. The polymerization process was conducted by incorporating 6 wt.% acrylamide monomers, 0.8 wt.% methylene-bisacrylamide, and a few milligrams of α,α-azoisobutyronitrile initiator into this solution. Heating the mixture from 70 to 90° accelerates this process. The green gel was gradually dried to 400 at 0.5°C/min.

The resultant black powder was ground in an agate mortar. Then, the powders were calcined at 900°C for 14 h. After this, we pressed the resulting powder into a disk shape. The obtained pellets were annealed in a tube-type furnace under flowing oxygen gas at a temperature of 915°C, and carbonate was removed from the precursors.

Finally, the precursors were mixed with Tl_2O_3 in a ratio of 1.2:1 and I_2O_5 (1.6, 2.0, 2.4 wt.%) and ground using a desktop automatic mixer. Following the last grinding stage, the powder was compressed using a hydraulic press set to 450 MPa to create a disk that was 10 mm in diameter and 4 mm in thickness. The obtained samples were placed in a gold crucible in quartz tubes and evacuated to 10^{-3} Torr. The furnace temperature was increased at a rate of 5°C/min to 900°C and maintained for 2 h, after which the samples were rapidly cooled at room temperature.

Using CuK_{α} radiation, X-ray powder diffraction patterns (XRD) were acquired using a Dron-3+PC. An energy dispersive spectrometer (EDS) from Oxford Instruments, AZtecOne, and the FEI Quanta FEG 600 SEM were used to examine the morphology and structure. Using the KBr disk technique, the samples' FT-IR spectrum was obtained within the 400–4000 cm^{-1} range on an FT-IR spectrometer (Cary 600 series). The resolution of the scan was 0.5 cm^{-1}. The samples were ground into a fine powder and combined 1:100 by weight with potassium bromide powder. The IR absorption spectra were taken as soon as the disks were ready.

We applied the high harmonics method to determine the critical temperature (T_c) and the transport critical current density, J_c. When the signal to be measured is less than 0.2 μV, the measurement error of high harmonics was about 2%; when the signal was higher, it was no more than 0.5%. The primary settings for the measurements were AC fields $h = 1$ Oe, frequency $f = 20$ kHz, and DC fields H = 0. Permalloy screens shield the Earth's magnetic field to less than 10^{-3} Oe.[22–24]

2.3 RESULTS AND DISCUSSION

Thermal annealing at the partial pressure of oxygen is the most effective way to produce precursors with ideal properties, particularly when the precursors are made from carbonates containing oxides. Precursor synthesis under oxygen pressure produces homogeneous cations, oxygen content, and removes carbonates from samples. Figure 2.1 displays the absorption spectra for the 400–4000 cm^{1} range of the precursors that were annealed at 900°C. The identification of carbonate within a system using X-ray phase analysis proves to be challenging due to its low sensitivity to this research technique. In this study, we investigated a precursor subjected to heating at 915°C degrees Celsius utilizing infrared spectroscopy. The absence of a fixed strong peak associated with the carbonyl group (C = O) in the range of 1400–1500 cm^{-1}, as observed in the figure, suggests that the thermal treatment effectively eliminated the carbonate from the system. This conclusion is further supported by the findings of scanning tunneling microscope analysis, which confirmed the presence of oxygen, barium, calcium, and copper in the precursor.

Figure 2.2 presents the results of the $Ba_2Ca_2Cu_3O_x$ precursor analyzed by SEM and EDS regarding the agglomerates' size and the sample's chemical composition. The average grain size is 0.2–3 μm (Figure 2.2a). This measurement was made on precursor, where temperature treatment was made at 915°C with flowing oxygen gas.

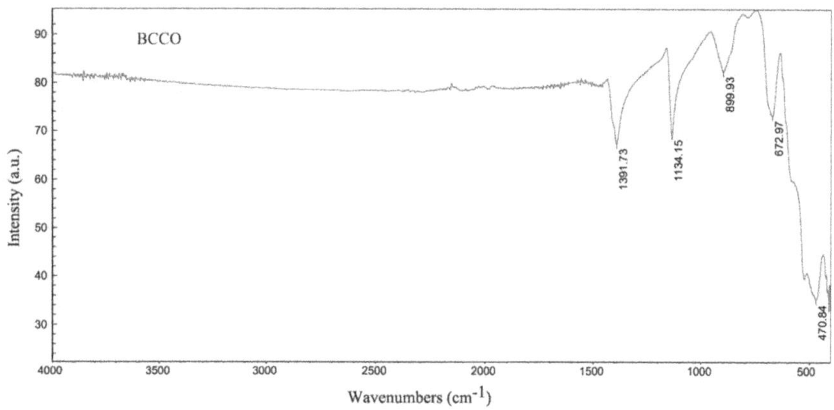

FIGURE 2.1 $Ba_2Ca_2Cu_3O_y$ infrared analysis following annealing at 915°C and 0.5 bar of flowing oxygen partial pressure.

Figure 2.2b presents an analysis of the chemical composition of $Ba_2Ca_2Cu_3O_x$ precursor. We revealed considerable amounts of barium and copper, which is consistent with the presence of the $BaCuO_2$ phase, and this analysis also indicates a substantial accumulation of Ca and Cu due to the formation of Ca_2CuO_3.

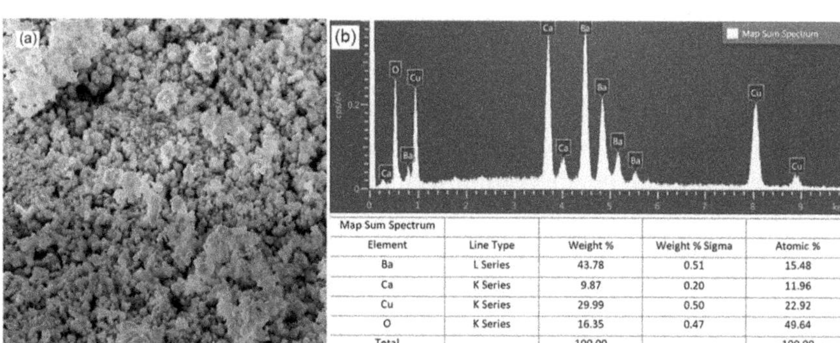

Map Sum Spectrum				
Element	Line Type	Weight %	Weight % Sigma	Atomic %
Ba	L Series	43.78	0.51	15.48
Ca	K Series	9.87	0.20	11.96
Cu	K Series	29.99	0.50	22.92
O	K Series	16.35	0.47	49.64
Total		100.00		100.00

FIGURE 2.2 SEM microphotograph of the $Ba_2Ca_2Cu_3O_x$ precursor. The spectrum of point analysis of EDS considers the composition of synthesized materials.

Figure 2.3 presents the X-ray diffraction (XRD) patterns of the $Ba_2Ca_2Cu_3O_y$ precursor material. We obtained only two $BaCuO_2$ and Ca_2CuO_3 phases, which agrees with the so-called pdf phase diagram of the $Ba_2Ca_2Cu_3O_y$ system.[30–32]

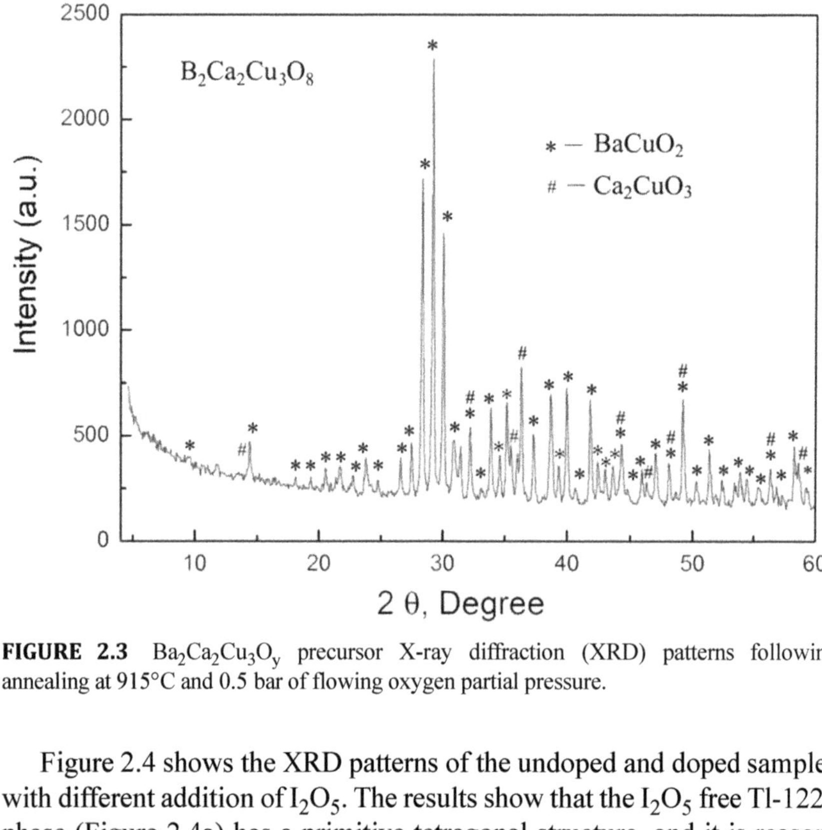

FIGURE 2.3 $Ba_2Ca_2Cu_3O_y$ precursor X-ray diffraction (XRD) patterns following annealing at 915°C and 0.5 bar of flowing oxygen partial pressure.

Figure 2.4 shows the XRD patterns of the undoped and doped samples with different addition of I_2O_5. The results show that the I_2O_5 free Tl-1223 phase (Figure 2.4a) has a primitive tetragonal structure, and it is reasonably consistent with the PDF faille with the lattice parameters $a = 0.31042$ and $c = 1.59413$ nm.

From Figure 2.4a, we can observe that the diffraction pattern of the undoped sample consists of unreacted $BaCuO_2$ precursor phases. In the cases of 1.6 wt.% dopant (I_2O_5) concentration (Figure 2.3b), the $BaCuO_2$ phase decreases. Increasing the dopant concentration until 2.0 wt.% (Figure 2.3c) leads to nearly a single Tl-1223 phase, and only two peaks of low intensity of the Tl-1212 phase are observed. However, in higher doping concentrations (2.4 wt.%), the impurity phase, such as Tl-1212 and BaCuO2, increases (Figure 2.4d), causing the superconducting phase to decrease.

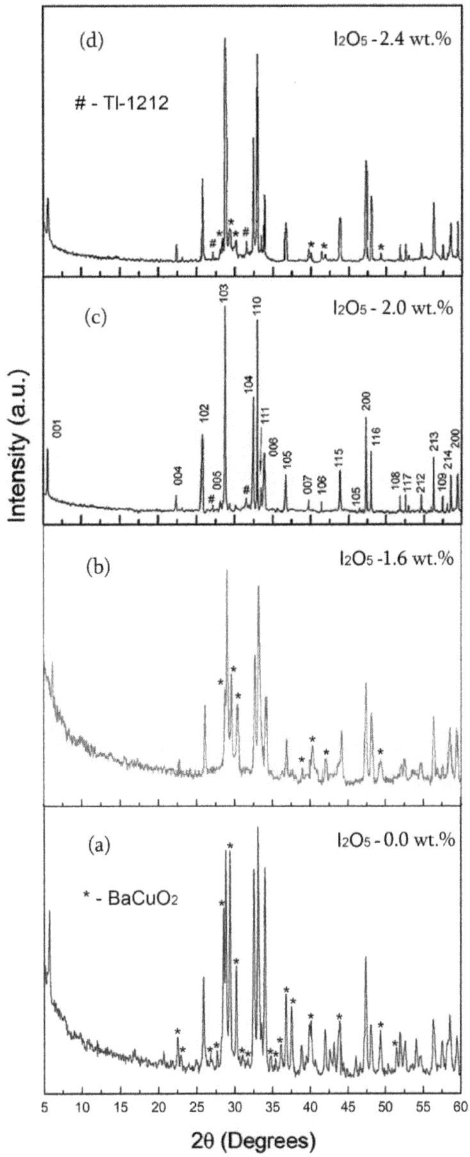

FIGURE 2.4 XRD patterns of the synthesized samples are doped with various quantities of I_2O_3 and iodine pentoxide-free samples.

Figure 2.5 ($\chi^3 = V_3$/h) displays measurements of the temperature dependence of the magnitude of the third harmonics module at $h = 1$ Oe, $f = 20$ kHz, and $H = 0$. The I_2O_5-free sample's superconducting transition begins at approximately $T_c = 112$ K, whereas the I_2O_5-doped sample exhibits $T_c = 115$ K. We found that the critical temperature, $T_c = 118$ K, reached 2.0%. The starting temperature drops to 108 K following the dopant enhancement at 2.4 wt.%. The plot of $|\chi^3|(T)$ shows two peaks. The low-temperature maximum is caused by penetration only into the Josephson medium, formed by weak links between the crystallites. In contrast, the high-temperature peak can be caused by penetration of the field into the crystallites and is associated with intragranular critical current density, j_{cg}. Intergranular critical current density, j_{cJ}, is linked to the latter.

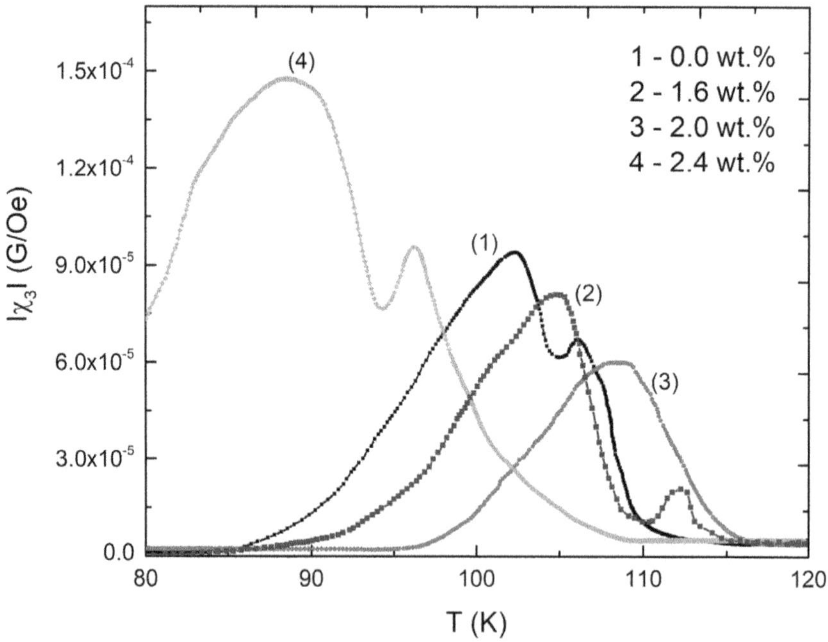

FIGURE 2.5 Temperature depends on the amplitude of the module of the third harmonics $|\chi_3|$.

Figure 2.6 presents the values of transport critical current density J_c vs temperature for I_2O_5-free and I_2O_5-doped Tl-1223 samples obtained from the $\chi3(T)$ data, $V_3 = 1/J_c$.

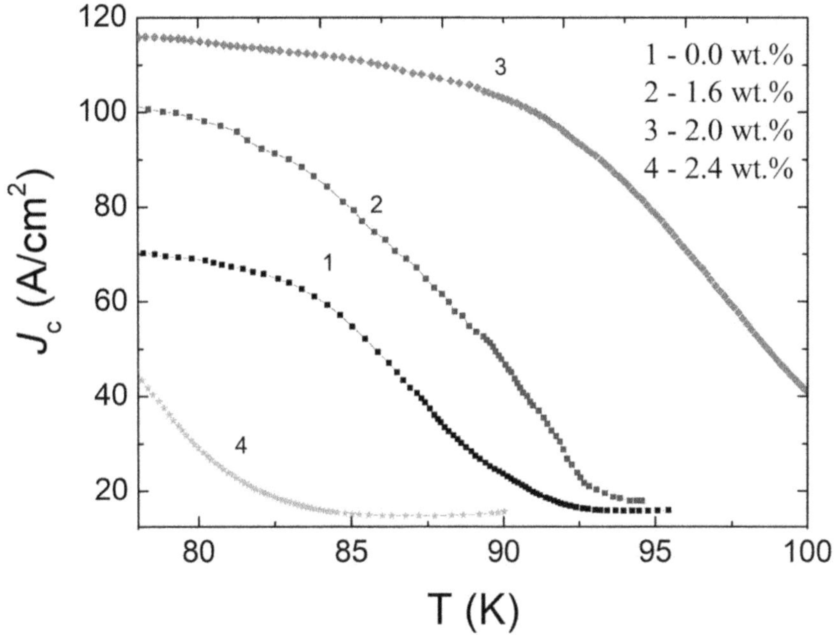

FIGURE 2.6 Transport critical current density as a function of temperature for Tl-1223 compounds, measured at $h = 1$ Oe, $f = 20$ kHz, and $H = 0$.

From Figure 2.6, we can see that the transport critical current density J_c for the I_2O_5-doped sample (2.0 wt.%) increases more than that for the I_2O_5-free sample. The J_c value for the un-doped sample is approximately 70 A/cm^2, and the largest value of critical currents is observed in sample 2.0 wt.%, which is equal to 115 A/cm^2.

2.4 CONCLUSION

In summary, in the present study, we synthesized a multiphase precursor powder with high chemical homogeneity and higher reactivity ability by the *in situ* polymerization method. We also examined the influence of different values of I_2O_5 on the thallination of the BCCO precursor.

Our experiments showed that the undoped sample observed more quantities of precursor $BaCuO_2$ phase, respectively, which influences critical temperature ($T_c \approx 112$ K) and critical current densities $J_c \approx 70$

A/cm². A low concentration of I_2O_3 (1.6 wt.%) also formed a nonpure Tl-1223 phase with a comparatively small amount of containing precursor $BaCuO_2$ phase, with the onset T_c at 115 K and transport critical current density at 100 A/cm². Unlike this, at 2.0 wt.% formed Tl-1223 near the single phase consists of a small number of impurity phases with critical temperature and current density 118 K and 115 A/cm², respectively—enhancement of I_2O_5 concentration at 2.4 wt.% leads to the degradation of Tl-1223 phases, with $T_c \approx 112$ K and $J_c \approx 70$ A/cm².

Finally, we conclude that only a 2.0 wt.% level of doped samples with I_2O_5 enhances the value of the transport critical current densities and critical temperature.

KEYWORDS

- **Tl-based superconductors**
- ***in situ* polymerization**
- **iodine pentoxide**
- **critical temperature (T_c)**
- **critical current density (J_c)**

REFERENCES

1. Xie, W.; Wang, P.; He, M.; Qiao, R.; Du, J.; Gao, X.; Liu, X.; Zhang, X.; Ji, L.; Chen, H.; Zhao, X. Double-sided $Tl_2Ba_2Ca_2Cu_3O_{10}$ Thin Films Based High Temperature Superconducting Filter Operating above 100K. *Appl. Phys. Lett.* **2014,** *105,* 092601.
2. Zhao, S.; Tian, W.; Zhang, X.; Liang, X.; He, Z.; Qi, Z.; Wang, P.; Ji, L.; He, M.; Wang, H. $Tl_2Ba_2CaCu_2O_8$ Bicrystal Josephson Junctions Operated Close to 100 K. *Supercond. Sci. Technol.* **2020,** *33,* 075006.
3. Bramley, A. P.; O'Connor, J. D.; Grovenor, C. R. M. Thallium-based HTS Thin Films, Processing, Properties and Applications. *Supercond. Sci. Technol.* **1999,** *12,* R57–R74.
4. Guo, R.; Duan, S.; He, Z.; Liang, X.; Niu, Z.; He, M.; Jiang, Y.; Wu, J.; Ji, L.; Jin, B.; Chen, J.; Wang, H.; Wu, P. Josephson Plasmon Resonance in $Tl_2Ba_2CaCu_2O_8$ High-Temperature Superconductor Tunable Terahertz Metamaterials. *Adv. Funct. Mater.* **2021,** *31,* 2106891.
5. Lauder, A.; Myers, K.; Face, D. Thin-Film High-Temperature Superconductors for Advanced Communications and Electronics. *Adv. Mater.* **1999,** *10,* 1249–1254.

6. Mikusu, S.; Takami, S.; Tokiwa, K.; Takeuchi, K.; Iyo, A.; Tanaka, Y.; Watanabe, T.; Neutron Powder Diffraction of the Superconductor $TlBa_2Ca_2Cu_3O_{8+\delta}$ with Different Maximum T_C values (T_C = 120–132 K). *Supercond. Sci. Technol.* **2008,** *21,* 085014.
7. Shipra, R.; Idrobo, J. C.; Sefat, A. S. Structural and Superconducting Features of Tl-1223 Prepared at Ambient Pressure. *Supercond. Sci. Technol.* **2015,** *28,* 115006.
8. Iyo, A.; Aizawa, Y.; Tanaka, Y.; Tokumoto, M.; Tokiwa, K.; Watanabe, T.; Ihara, H. High-pressure Synthesis of $TlBa_2Ca_{n-1}Cu_nO_y$ (n = 3 and 4) with T_c=133.5 K (n = 3) and 127 K (n = 4). *Physica C* **2001,** *357–360,* 324–328.
9. Iyo, A.; Tanaka, Y.; Ishiura, Y.; Tokumoto, M.; Tokiwa, K.; Watanabe, T.; Ihara, H.; Study on Enhancement of *Tc* (≥130 K) in $TlBa_2Ca_2Cu_3Oy$ Superconductors. *Supercond. Sci. Technol.* **2001,** *14,* 504–510.
10. Liu, X. Preparation and Properties of YBCO/Bi-2212/YBCO Superconducting Trilayer Films by Acetate Based Sol–gel Method. *J. Sol-Gel. Sci. Technol.* **2021,** *98,* 68–75.
11. Metskhvarishvili, I. R.; Lobzhanidze, T. E.; Dgebuadze, G. N.; Bendeliani, B. G.; Metskhvarishvili, M. R.; Rusia, M. S. H.; Giorganashvili, G. R.; Gabunia, V. M. Comparative study of Tl-1223 Superconductors Prepared by the Sol-gel Route and Solid-state reaction. *Low Temp. Phys.* **2022,** *48,* 3–6.
12. Sin, A.; Odier, P.; Nuez-Regueiro M.; Sol–gel Processing of Precursor for High-T Superconductors: c Influence of Rhenium on the Synthesis of $Ba_2Ca_2Cu_3O_x$. *Physica C* **2000,** *330,* 9–18.
13. Mendonca, T. M.; Tavares, P. B.; Correia, J. G.; Lopes, A. M. L.; Darie, C.; Araujo, J. P. The Urea Combustion Method in the Preparation of Precursors for High-T_C Single Phase $HgBa_2Ca_2Cu_3O_{8+\delta}$ Superconductors. *Physica C* **2011,** *471,* 1643–1646.
14. Brylewski, T.; Przybylski, K.; Morawski, A.; Gajda, D.; Cetner, T.; Chmis, J. The Urea Combustion Method in the Preparation of Precursors for High-TC Single Phase $HgBa_2Ca_2Cu_3O_{8+d}$ Superconductors. *J. Adv. Cera.* **2016,** *5,* 185–196.
15. Khan, N. A.; Nadeem, A. M.; Safeer, S. H. Enhancement of Inter-plane Coupling and Superconductivity in $Cu_{0.5}Tl_{0.5}Ba_2(Ca_{3-y}M_y)Cu_4O_{12-\delta}$ (M = 0, Pr, Y, Er, Gd; y = 0.02) Samples *J. Supercond. Nov. Magn.* **2016,** *28,* 2253–2260.
16. Shipra, R.; Sefat, A. S. Effect of Li_2O on the Microstructure, Magnetic and Transport Properties of Tl-2223 Superconductor. *Physica C* **2015,** *519,* 108–111.
17. Jasim, K. A. The Effect of Cadmium Substitution on the Superconducting Properties of $Tl_{1-x}Cd_xBa_2Ca_2Cu_3O_{9-\delta}$ Compound. *J. Supercond. Nov. Magn.* **2013,** *26,* 549–552.
18. Mohammed, N. H. The Excess Conductivity of $(Cu_{0.5}Tl_{0.5})$-1223 Superconductor Substituted by Ti. *Physica C* **2013,** *485,* 95–101.
19. Khurram, A. A.; Khan, N. A.; Ahadian, M. M.; Iraji-zad, A. *Physica C* **2008,** *468,* 405–410.
20. Mumtaz, M.; Ali, L.; Ahmad, I. Phase Formation, Activation Energy and Superconductivity of MgO Nanoparticles Added $(Cu_{0.5}Tl_{0.5})Ba_2Ca_2Cu_3O_{10-\delta}$ Phase. *Physica C* **2018,** *551,* 19–24.
21. Elokr, M. M.; Awad, R.; El-Ghany, A. A. Effect of Nano-Sized ZnO on the Physical Properties of $(Cu_{0.5}Tl_{0.25}Pb_{0.25})Ba_2Ca_2Cu_3O_{10-\delta}$. *J. Supercond. Nov. Magn.* **2011,** *24,* 1345–1352.

22. Metskhvarishvili, I. R.; Dgebuadze, G. N.; Bendeliani, B. G.; Lobzhanidze, T. E.; Metskhvarishvili, M. R.; Mumladze, G. N. Low ac Field Response of Bi-based Superconductors with Addition of Antimony Oxide. *J. Phys.: Conf. Ser.* **2014,** *507,* 012032.
23. Metskhvarishvili, I. R.; Dgebuadze, G. N.; Bendeliani, B. G.; Metskhvarishvili, M. R.; Lobzhanidze, T. E.; Mumladze, G. N.; Low Field ac Susceptibility and High Harmonics Studies in $PbMo_6S_8$ Polycrystalline Superconductor. *J. Low Temp. Phys.* **2013,** *170,* 68–74.
24. Metskhvarishvili, I. R.; Bendeliani, B. G.; Dgebuadze, G. N.; Giorganashvili, G. R.; Metskhvarishvili, M. R.; Lobzhanidze, T. E. Low-field High-harmonic Generation in $Mo_6S_6I_2$ Chevrel-phase Superconductor. *Low Temp. Phys.* **2022,** *48,* 16–19.
25. Bean, C. P. Magnetization of High-Field Superconductors. *Rev. Mod. Phys.* **1964,** *36,* 31–34.
26. Fukumoto, Y.; Matsushita, T. The Effect of Reversible Flux Motion on the Estimation of Critical Current Density in a Thin Superconductor using a Third Harmonic Voltage Method. *Supercond. Sci. Technol.* **2005,** *18,* 861.
27. Mancusi, D.; Galluzzi, A.; Pace, S.; Polichetti, M. Effects of the Third Harmonic Demagnetization Field on the ac Losses of a Granular FeSeTe Superconductor in ac Field. *J. Supercond. Nov. Magn.* **2018,** *31,* 2011–2018.
28. Gioacchino, D. D.; Puri, A.; Marcelli, A.; Saini, N. Flux Dynamics Iron-Based Superconductors. *IEEE Trans. Appl. Supercond.* **2013,** *23,* 7300505.
29. Mihalache, V.; Popa, S.; Gioacchino, D. D.; Tripodi, P.; Vinko, J. D. Experimental Evidence for a Dimensional Crossover of the Vortex Ensemble in BSCCO ($Bi_{1.8}Pb_{0.4}Sr_{1.8}Ba_{0.2}Ca_2Cu_3O_x$) by Multi-harmonic ac Susceptibility Measurements. *Physica C.* **2006,** *433,* 225–233.
30. Wong-Ng, W.; McMurdie, H.; Paretzkin, B.; Zhang, Y.; Davis, K.; Hubbard, C.; Dragoo, A.L.; Stewart, J. M. Reference X-Ray Diffraction Powder Patterns of Fifteen Ceramic Phases. *Powder Diffr.* **1987,** *2,* 257–265.
31. Wong-Ng, W.; Cook, L. A Review of the Crystallography and Crystal Chemistry of Compounds in the BaO-CuO_x System. *Powder Diffr.* **1994,** *9,* 280–289.
32. Whiter, J. D.; Roth, R.S. *Phase Diagrams for High-Tc Superconductors*; Amer. Ceramic Society, Westerville, Ohio, USA, 1991, p 170.

CHAPTER 3

Synthesis of a Novel Hybrid (Bis-Azo Group Containing Spirochromene) Photochromic Compound

TINATIN BUKIA[1,2,3], LUKA BUSKANDZE[2], NODAR BERIDZE[2], ANA GOLETIANI[3], TAMAR TATRISHVILI[3,4], and GIA PETRIASHVILI[1]

[1]*V. Chavchanidze Institute of Cybernetics, Georgian Technical University, Tbilisi, Georgia*

[2]*Scientific-Research Institute of Organic Chemistry Faculty of Exact and Natural Sciences Iv. Javakhishvili Tbilisi State University, Tbilisi, Georgia*

[3]*Institute of Macromolecular Chemistry and Polymeric Materials, Ivane Javakhishvili Tbilisi State University, Tbilisi, Georgia*

[4]*Ivane Javakhishvili' Tbilisi State University, Department of Macromolecular Chemistry, Tbilisi, Georgia*

ABSTRACT

The synthesis of a novel bis-azo group containing spirochromene was carried out in two ways: in one case, the reaction was carried out by reduction of 1,2,3,3-tetramethyl-5-nitro-3*H*-indol-1-ium iodide in the presence of $NaBH_4$/$LiAlH_4$ and subsequent interaction of the obtained compound with 3,5-dinitro salicylaldehyde, and in the second case, the synthesis was carried out by reduction of 5-nitro-2,3,3-trimethylindolenine with $NaBH_4$/$LiAlH_4$ to give 1,2-bis(2,3,3-trimethylindolin-5-yl)diazene, which was then oxidized with MnO_4 to give the bis-azo compound. By quaternization of the obtained bis-azo compound, (*Z*)-5,5′-(diazene-1,2-diyl)bis(1,2,3,3-tetramethyl-3*H*-indol-1-ium)iodide was

Advances in Engineering Materials: Preparation for Sustainable Process Development. Tamara Tatrishvili and A. K. Haghi (Eds.)

formed, which was further treated with alkali and reacted with 3,5-dinitro salicylaldehyde to give bis-azo group containing spirochromene. The structures of the synthesized compounds were determined by physicochemical analysis.

3.1 INTRODUCTION

Azo compounds are the most versatile derivatives in organic chemistry. They have gained considerable attention in the last decade for their various applications in the fields of cosmetology, the clothing industry, biomedicine, optical systems, photochromic devices, molecular photoswitches, etc.[1–10]

The potential of azo compounds for the creation of the next generation of intelligent bio-nanomaterials has been highlighted by their rapid and reversible responses to a variety of external stimuli, including temperature, pH, precise light wavelengths, coordination with anions or cations, and the distinct properties of their isomers.

The capability of azo compounds in responding to several external stimuli, such as precise light wavelength, pH, temperature, and coordination with anions or cations, in a rapid and reversible way, as well as the different properties of their isomers, have evidenced the potentials of these systems for the development of the next generation of smart bionanomaterials. Azo compounds are organic complexes with the following molecular formula Ar−N=N−Ar′, which contain two nitrogen atoms connected by a double bond. The nature of the aromatic substituents on either side of the azo group controls the colors and properties of the azo compounds. Depending on the exact structure of the molecule, they are typically brightly colored compounds that can be intensely yellow, red, orange, blue, or even green.[4] Also, azo dyes have been extensively researched for their superior thermal and optical properties in applications such as oil-soluble lightfast dyes, toners, optical recording media, and inkjet printing. Recently, azo compounds have also attracted attention as organic dyes due to their intriguing electrical properties in photoconductors.[11–17] Because spirochromenes have a wide range of applications, it is important to understand their underlying chemistry. To achieve this goal, spirochromenes containing azo dyes were synthesized and studied. Chemists, physicists, and engineers paid close attention to the unique physical and chemical properties of spiropyrans when Wizinger and Wenning[5–10] reported them in 1940. Molecular ensembles and bistable molecules can exist in two different thermodynamically stable states. Different stimuli can induce a change in state, including light, heat, mechanical stress, electric and magnetic fields, etc. An uncolored spiropyran molecule with a bulk structure transforms into a coplanar-colored merocyanine form with a high dipole moment when exposed to ultraviolet (UV) radiation. Highly sensitive and selective chemosensors are being developed based on spiropyrans. Light can control the permeability of natural and artificial membranes, as well as the activity of enzymes, thanks to spiropyrans! The ability of an open (colored) spiropyran molecule to form micelles due to the presence of a long fatty radical at the nitrogen atom is known as surface activity.[2–4,30] This property facilitates the transfer of biological information. Spiropyrans are used as molecular switches because of their high quantum yields of photoinduced regrouping and striking spectral and geometric contrasts of mutually transformable isomeric forms.

The spiropyran and azo dyes are bound together in an unexposed form, and the two dyes (azo and merocyanine) develop symbiotically when exposed to UV light. The key is that the molecule has electron-deficient substituents at one end and electron-donating substituents at the other. The visible spectral region of the compound will have an azo dye spectrum prior to exposure that follows the nature of the substituent(s), and the linked, flat, conjugated chain spectrum of the two dyes will appear after exposure. This work was mainly oriented to the synthesis of bis-azo group containing spirochromene based on 2,3,3-trimethyl-5-nitro-3*H*-indolenine according to the following Schemes 3.1 and 3.2:

3.2 EXPERIMENTAL METHODS AND MATERIALS

All reagents used for the synthesis were purchased from commercial suppliers from Sigma–Aldrich.

Fourier transform infrared spectra (FT-IR) were recorded at Nicolet TM iS50 FTIR Spectrometer-Thermo Fisher Scientific, the infrared (IR) region 4000–400 cm^{-1} (scan 32, resolution 4 cm^{-1}); UV spectra were recorded on GENESYS 50 spectrophotometer Thermo Fisher Scientific.

^1H and ^{13}C proton nuclear magnetic resonance (NMR) spectra were recorded on a Bruker NMR-400 (400 MHz) in DMSO-d_6 or $CDCl_3$ with TMS as the internal standard. Mass spectra were recorded on Agilent Technologies 6539 Accurate-Mass Q-TOF LC–MS. Monitoring of the reaction progress and assessment of the purity of the synthesized compounds was performed by thin layer chromatography (TLC) on silica gel on TLC PET-foil plates in hexane/ethyl acetate, 2:1 eluent system, visualization in an iodine chamber and UV light. Melting points were determined using a Stuart SM10G.

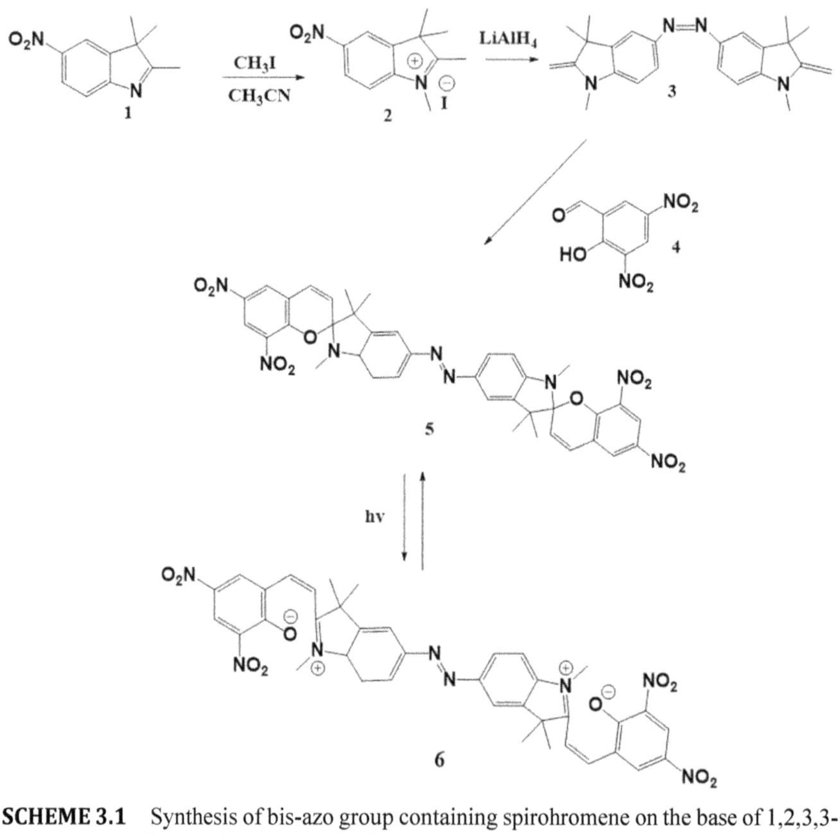

SCHEME 3.1 Synthesis of bis-azo group containing spirohromene on the base of 1,2,3,3-tetramethyl-5-nitro-3*H*-indol-1-ium iodide.

3.2.1 EXPERIMENTAL PART

1,2,3,3-tetramethyl-5-nitro-3*H*-indol-1-ium iodide (2). 5 g (0.0244 mol) of compound **1** in 10 mL acetonitrile was added to 7.62 mL (0.122 mol) of CH_3I. The quaternization was performed in a sealed ampoule at 80°C for 24 h. Formed crystals were washed with diethyl ether, dried, and recrystallized from methanol. Compound 3, as dark yellow crystals with 90% yield, was isolated.

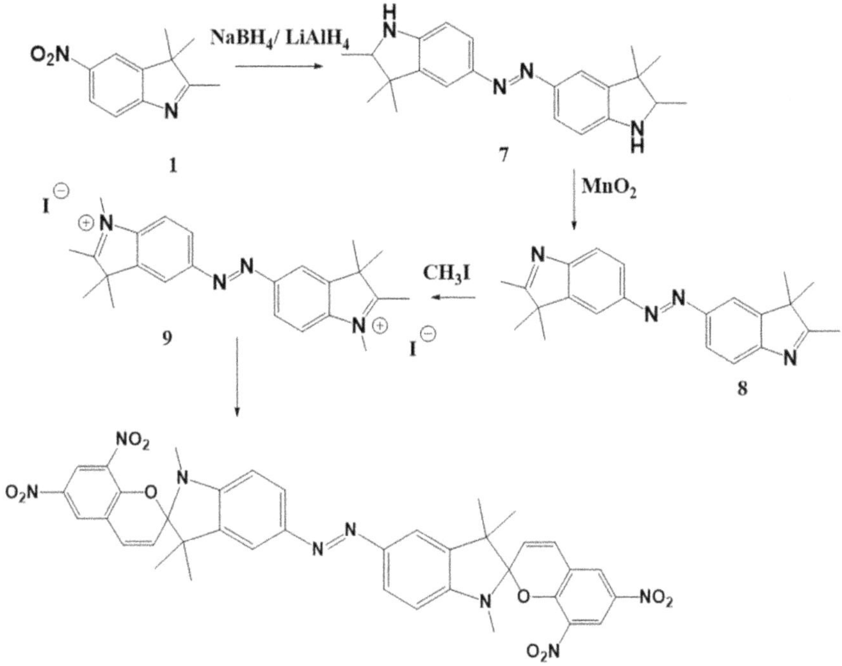

SCHEME 3.2 Synthesis of bis-azo group containing spirohromene on the base of 5-nitro-2,3,3,-trimethyl-3*H*-indolenine.

IR spectrum, wavenumber, cm^{-1}: 2963, 2935, 2846 (methyl C–H asym/sym. Stretch) 1586,1375 (NO_2), 1344 (aromatic tertiary amine C–N stretch), 3050 (C–H aromatic ring stretch); ^{1}H NMR (400 MHz, D_2O, δ ppm): 1.15 (3H, s, CH_3) 1.45 (6H, s, CH_3), 7.06–7.08 (1H, d, j = 8.6 Hz, Ar), 8.09 (1H, s, Ar), 8.13–8.15 (1H, dd, j = 8.7, 2.3 Hz, Ar); ^{13}C NMR (100 MHz, $CDCl_3$ δ ppm): 21.8 (2C, CH_3); 24.0 (1C, CH_3); 27.0 (1C), 42 (1C), 116.0 (1C, Ar); 123.4 (1C, Ar); 131.5 (1C, Ar); 146.6 (2C, Ar); 151 (1C, Ar); 171 (1C, Ar);

(*Z*)-1-((*R*)-1′,3′,3′-trimethyl-6,8-dinitrospiro[chromene-2,2′-indolin]-5′-yl)-2-((S)-1′,3′,3′-trimethyl-6,8-dinitrospiro[chromene-2,2′-indolin]-5′-yl)diazene (5) In 100-mL flask equipped with a magnetic stirrer was placed 2 g (5.78 m.mol) of

1,2,3,3-tetramethyl-5-nitro-3*H*-indol-1-ium iodide (2) in 50 mL of methanol. To the suspension, at room temperature, 1.09 g (28.9 mmoll) of $NaBH_4$ was added with a small portion. During the addition of $NaBH_4$, the reaction mixture starts changing its color, and it gets transparent yellow. The reaction was monitored by TLC. After finishing the reaction, it was transferred into a separatory funnel, added 100 mL of water, and the aqueous layer was extracted 3 times with 50 mL of diethyl ether. Ether extracts were dried over KOH, concentrated on the vacuum, and obtained 1.48 g (4.1 mmol, 71%) compound 3 as a brownish oil. 0.87 g (4.1 mol) of 3,5-dinitro-salicylaldehyde dissolved in 5 mL of alcohol was added to the solution of the obtained compound 3. The reaction mixture was stirred and heated for 2 h. At the end of the reaction, the precipitate formed was filtered and recrystallized from methanol. The spiro compound 5 was obtained in 55% yield (1.7 g).

IR spectrum, wavenumber, cm^{-1}: 2965, 2931, 2853 (methyl C–H asym/sym. stretch) 1585–1475 (open-chain azo (–N=N–)), 1553, 1375 (NO_2), 1345, 1343 (aromatic tertiary amine C–N stretch), 3050 (C–H aromatic ring stretch); ^1H NMR (400 MHz, $CDCl_3$, δ ppm): 1.37 (6H, s, CH_3), 1.48 (3H, s, CH_3), 6.05–6.07(1H, d, j = 8.6 Hz), 6.7–6.8 (1H, d, j = 2.3 Hz), 7.14–7.16 (1H, dd, j = 8.7, 2.3 Hz, Ar), 7.84–7.86 (1H, dd, j = 8.7, 2.3 Hz, Ar), 7.96–7.98 (1H, dd, j = 8.7, 2.3 Hz, Ar), 8.12–8.14 (1H, dd, j = 8.7, 2.3 Hz, Ar), 8.31 (1H, s, Ar); ^{13}C NMR (100 MHz, $CDCl_3$ δ ppm): 21.6 (2C, CH_3); 25.0 (1C, CH_3); 51 (1C), 117.8 (1C, Ar); 118.2 (2C, Ar); 122.5 (4C, Ar); 138.7 (2C, Ar); 141 (2C, Ar); 145 (2C, Ar); 144 (1C, Ar); 148 (1C, Ar); LC–MS, m/z (Irel, %): found 761.112 [M + 1] (100%), calculated *m/z*: 760.720.

1,2-bis(2,3,3-trimethylindolin-5-yl)diazene (7) method 1. In a 100-mL flask equipped with a magnetic stirrer was placed 5 g (0.024 mol) of 2,3,3-trimethyl-5-nitro-3*H*-indolenine in 50 mL of methanol. To the clear solution, 4.5 g (0.12 mL) of $NaBH_4$ was added with a small amount at room temperature. During the addition of $NaBH_4$, the reaction mixture starts to change its color and becomes yellowish. The reaction was monitored by TLC. After completion of the reaction, the mixture was transferred to a separating funnel, 100 mL of water was added, and the aqueous layer was extracted 3 times with 50 mL of diethyl ether. The ether extracts were dried over KOH and then concentrated under vacuum to give the yellowish compound 7 in high yield (99%). IR spectrum, wavenumber, cm^{-1}: 2963, 2922, 2859 (Methyl C–H asym/sym. Stretch) 1585–1473 (open-chain azo (–N=N–)), 1374, 1340 (aromatic tertiary amine C–N stretch), 3050 (C–H Aromatic ring stretch); ^{1}H NMR (400 MHz, $CDCl_3$, δ ppm): 1.13 (6 H, s, CH_3) 1.36 (3H, s, CH_3), 2.99 (3H, s, CH_3), 5.65 (1H, s, NH), 7.04–7.06(1H, d, j = 8.6 Hz, Ar), 8.08–8.19 (1H, d, j = 2.3 Hz, Ar), 8.13–8.15 (1H, dd, j = 8.7, 2.3 Hz, Ar); ^{13}C NMR (100 MHz, $CDCl_3$ δ ppm): 19.6 (2C, CH_3); 24.0 (1C, CH_3); 42 (1C), 65.4 (1C), 113.9 (1C, Ar); 122.2 (1C, Ar); 123.5 (1C, Ar); 135.7 (2C, Ar); 139 (1C, Ar); 151 (1C, Ar); LC-MS, m/z (Irel, %): found 349.2124 [M + 1] (100 %), calculated *m*/*z*: 348.2314.

Method 2. 5 g (0.024 moll) of 2,3,3-trimethyl-5-nitro-3*H*-indolenine in 50 mL abs. dirthyl ether was added dropwise to the suspension of 4.56 g (0.12 moll) $LiAlH_4$ in diether solution. The reaction proceeds vigorously and ends within 2–5 min. It was worked up as method 1. The yield of the obtained compound **7** was 99 %.

(*Z*)-1,2-bis(2,3,3-trimethyl-3*H*-indol-5-yl)diazene (8). In a 100-mL three-necked flask equipped with a reflux condenser, thermometer, and heating mantle was placed 0.5 g (1.44 mmol) of 1,2-bis(2,3,3-trimethyl-indolin-5-yl)diazene (7) in 20 mL of toluene and was added 1.26 g. (14.5 mmol) of manganese(IV)dioxide. The mixture was refluxed for 15 h. with water removal (Dean-Stark), and then the formed suspension was filtered and washed several times with toluene. The filtrate was evaporated in a vacuum to give a dark orange solid **8**. Obtained reddish-orange powder with 45% yield. It was recrystallized from methanol.

IR spectrum, wavenumber, cm^{-1}: 2961, 2925, 2858 (methyl C–H asym/sym. stretch) 1586–1473 (open-chain azo (–N=N–)), 1376, 1344 (aromatic tertiary amine C–N stretch), 3050 (C–H aromatic ring stretch); ^{1}H NMR (400 MHz, $CDCl_3$, δ ppm): 1.13 (3 H, s, CH_3) 1.37 (3H, s, CH_3), 1.48 (3H, s, CH_3), 7.05–7.07(1H, d, j = 8.6Hz, Ar), 8.09–8.10 (1H, d, j = 2.3 Hz, Ar), 8.14–8.16 (1H, dd, j = 8.7, 2.3 Hz, Ar); ^{13}C NMR (100 MHz, $CDCl_3$ δ ppm): 21.6 (2C, CH_3); 25.0 (1C, CH3); 42(1C), 118.9 (1C, Ar); 125.2 (1C, Ar); 142.5 (1C, Ar); 148.7 (2C, Ar); 159 (1C, Ar); 165 (1C, Ar); LC–MS, *m*/*z* (Irel, %): found 344.3524 [M + 1] (100%), calculated *m*/*z*: 344.2001.

(*Z*)-5,5′-(diazene-1,2-diyl)bis(1,2,3,3-tetramethyl-3*H*-indol-1-ium) iodide (9) was synthesized by interaction of 2 g (5.8 mmol) of (*Z*)-1,2-bis(2,3,3-trimethyl-3*H*-indol-5-yl)diazene (10) with 4.1 g (1.8 mL, 29 mmol) of CH_3I in 10 mL acetonitrile. The quaternization was performed in a shallow ampoule for 24 h, heating at 80–82°C. After the finishing of the reaction obtained quaternized salt 9. It was worked up with diethyl ether, and the formed precipitate was filtrated and recrystallized from methanol. The yield of the obtained product **9** was 78%.

IR spectrum, wavenumber, cm^{-1}: 2961, 2925, 2858 (methyl C–H asym/sym. Stretch) 1586–1473 (open-chain azo (–N=N–)), 1376, 1344 (aromatic tertiary amine C–N stretch), 3050 (C–H aromatic ring stretch); ^1H NMR (400 MHz, $CDCl_3$, δ ppm): 1.13 (3 H, s, CH_3) 1.37 (3H, s, CH_3), 1.48 (3H, s, CH_3), 7.05–7.07(1H, d, j = 8.6 Hz, Ar), 8.09–8.10 (1H, d, j = 2.3 Hz, Ar), 8.14–8.16 (1H, dd, j = 8.7, 2.3 Hz, Ar); ^{13}C NMR (100 MHz, $CDCl_3$ δ ppm): 21.6 (2C, CH_3); 25.0 (1C, CH_3); 42(1C), 118.9 (1C, Ar); 125.2 (1C, Ar); 142.5 (1C, Ar); 148.7 (2C, Ar); 159 (1C, Ar); 165 (1C, Ar);

(Z)-1-((R)-1′,3′,3′-trimethyl-6,8-dinitrospiro[chromene-2,2′-indolin]-5′-yl)-2-((S)-1′,3′,3′-trimethyl-6,8-dinitrospiro[chromene-2,2′-indolin]-5′-yl)diazene(5) 1 g(1.6mmol)of(*Z*)-5,5′-(diazene-1,2-diyl) bis(1,2,3,3-tetramethyl-3*H*-indol.-1-ium) iodide (**9**) was dissolved in ethanol/water 1/1, added 5 mL of KOH 5%, and heated 45–50°C for 3 h. Obtained darkish residue was extracted 3 times with 30 mL of diethyl ether. The ether extract was dried over KOH and concentrated in a vacuum. The solution of 0.34 g (1.6 mmol) 3,5-dinitro-salicylaldehyde in 5 mL of alcohol was added to the alcohol solution of the obtained intermediate. The reaction mixture was stirred and heated for 2 h. After finishing the reaction, the formed precipitate was filtered and recrystallized from methanol; the spiro-compound **5** was obtained with a 55% yield.

3.3 RESULTS AND DISCUSSION

The synthesis of bis-azo group containing spirochromene was accomplished in two different pathways, shown in Schemes 3.1 and.2. According to Scheme 3.1, compound **1** was quaternized using CH_3I in acetonitrile. The obtained crystals were washed with diethyl ether and recrystallized from methanol. Reduction of compound **2** was carried out using an excess amount of NaBH4 or LiAlH4 in an absolute solvent. The reaction proceeds at room temperature with vigorous stirring. The color change from dark yellow to light yellow occurs within 3–5 min after adding the catalyst. After finishing the reaction, the mixture was transferred to a separatory funnel; treated with water; extracted with diethyl ether and dried over KOH plates; decanted into a round-bottom flask; and concentrated to a dry residue. TLC of the obtained compound showed that two compounds with different R_f were formed. The ^1H NMR and mass-spectral analytical study showed that together with compound **3**, it forms bis-azo compound **10** (Scheme 3.3), which does not form a spirochromene by heating with aldehydes. During the reduction of nitro compound **2** using both catalyze the nitro group is reduced to the –N=N– bond; also, the C=N– double bond is partially restored to the CH–NH bond, which is confirmed by the ^{1}HNMR and IR studies.

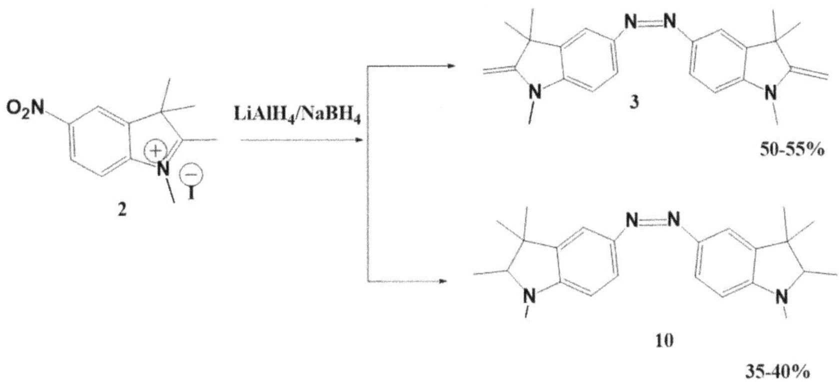

SCHEME 3.3 Catalytic reduction of 1,2,3,3-tetramethyl-5-nitro-3*H*-indol-1-ium iodide.

The same process was observed in the reduction of compound 1 using $NaBH_4/LiAlH_4$ at room temperature, and bis-azo compound 7 was obtained in high yield (Scheme 3.2). The research demonstrates that the diazo group in the molecule stays stable while the pyrrole ring in the indolenine nucleus is reduced during the interaction of $NaBH_4/LiAlH_4$ with 5-nitro-2,3,3,-trimethyl-3*H*-indolenine 1.

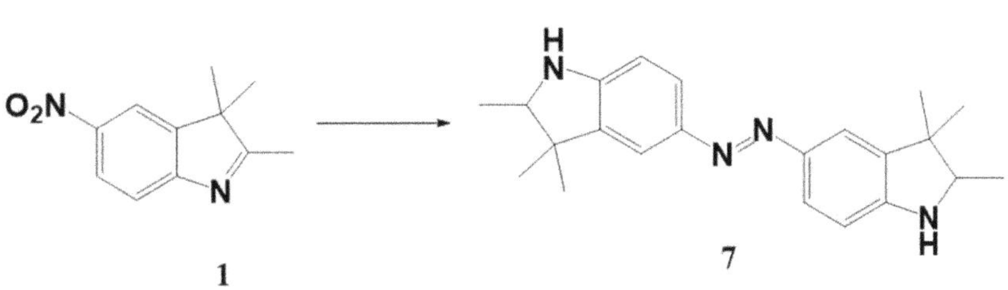

SCHEME 3.4 Catalytic reduction of 5-nitro-2,3,3,-trimethyl-3*H*-indolenine (**1**).

The IR spectra of compounds **1** and **7** manifest the characteristic absorption bands of aromatic rings at 3093 and 1599 cm^{-1}, respectively. For compound 1, the $C–NO_2$ absorption bands are located at the 1334 and

1514 cm^{-1} area (Figure 3.1), and absorption bands of NH are shown at 3358 cm^{-1} region for compound **7** (Figure 3.2). –N=N– groups absorption of compound **7** appeared at 1487 cm^{-1} region (Figure 3.2).

In ^{1}H NMR spectra of synthesized compound **7**, aromatic proton signals appear as singlets, doublets, doublets of doublets, and multiples depending on the influence of the respective substituents in the 8.96–6.95 ppm range. NH group signals appear as singlets at 4.97 ppm. Proton signals of the CH_3 group in the spectra appear as multiples in the upfield region of the spectra (Figure 3.3).

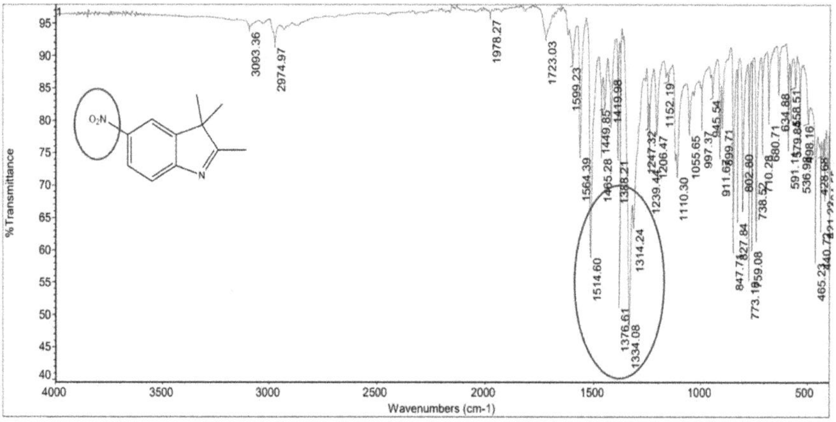

FIGURE 3.1 IR spectrum of 5-nitro-2,3,3-trimethyl indolenine (1).

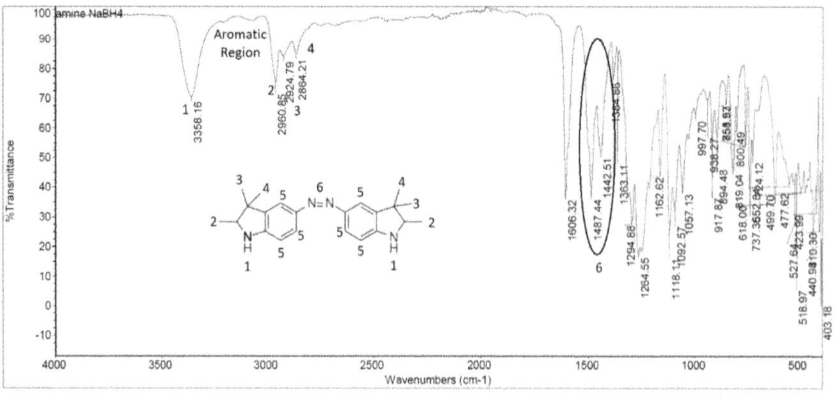

FIGURE 3.2 IR spectrum of 1,2-bis(2,3,3-trimethylindolin-5-yl)diazene (7).

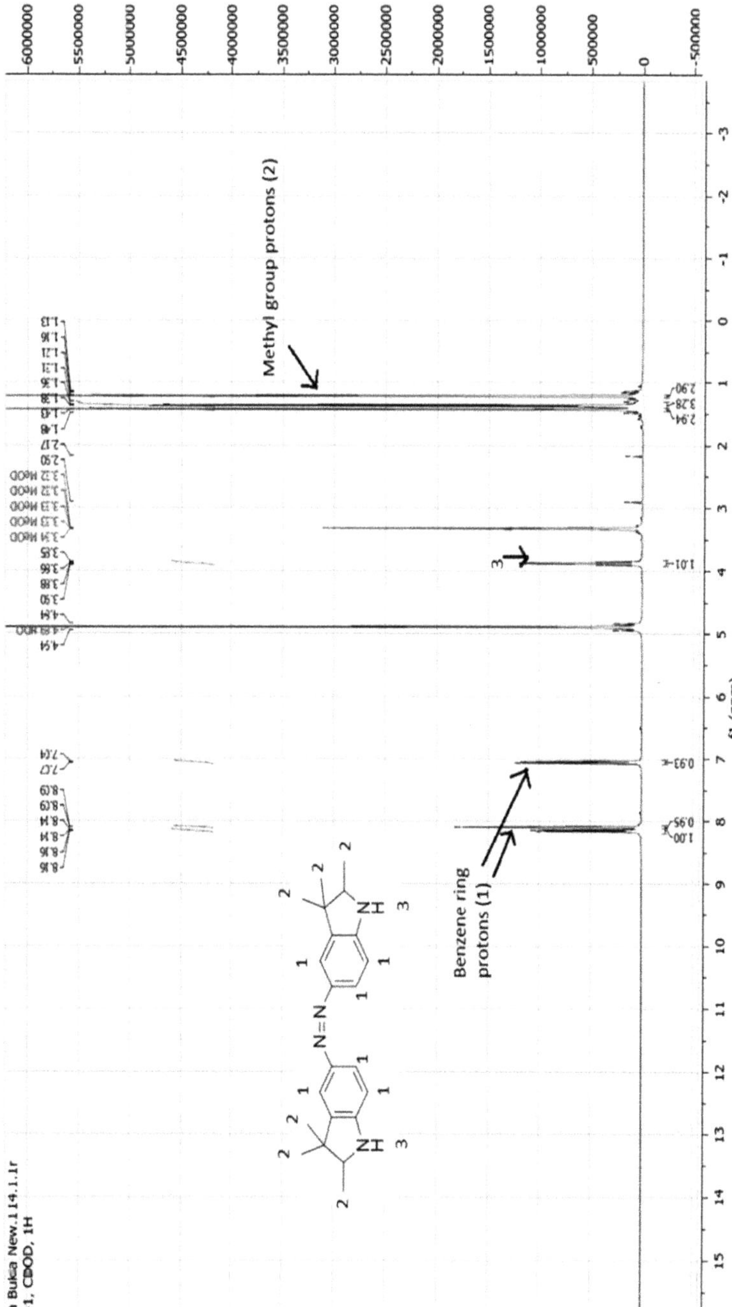

FIGURE 3.3 ^{1}H NMR spectrum of 1,2-bis(2,3,3-trimethylindolin-5-yl)diazene (9).

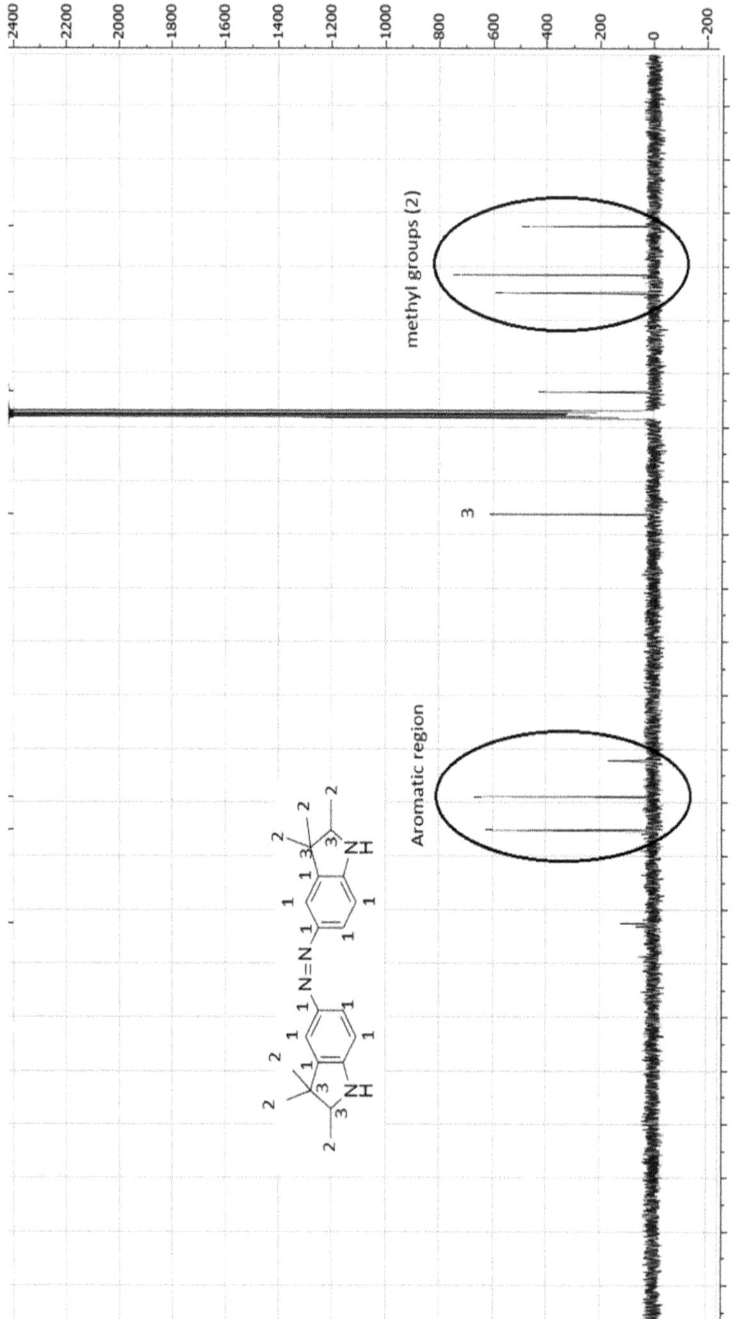

FIGURE 3.4 ^{13}C NMR spectrum of 1,2-bis(2,3,3-trimethylindolin-5-yl)diazene (9).

In ^{13}C NMR spectra of compound 3, the carbon atom signals of the CH_3 group appear in the 16.1–39.37 ppm upfield region. Resonance signals of the aromatic ring appear in the 109–165 ppm regions, respectively (Figure 3.4).

According to Scheme 3.1, a bis-azo group containing spirochromene **5** was obtained by interaction of compound 3 with 3,5-dinitrosalicylic aldehyde. The reaction was implemented in ethanol. from which a mixture of two products, compound **5** (50–55%) and compound **10** (35–40%), was obtained. Their separation was conducted by column chromatography. According to Scheme 3.2, the same compound 5 was obtained by alkaline treatment of quaternized bis-azo compound **9** and further interaction with 3,5-dinitrosalicylic aldehyde, where the yield of the final product 5 was 49%.

3.4 CONCLUSION

This research studied the catalytic reduction of 1,2,3,3-tetramethyl-5-nitro-3*H*-indol-1-ium iodide and 5-nitro-2,3,3-trimethyl-3*H*-indolenine using $NaBH_4$ and $LiAlH_4$ in an appropriate solvent.

Catalytic reduction of both compounds afforded the corresponding bis-azo compounds very high yields. Interaction of the obtained bis-azo compounds with 3,5-dinitrosalicylic aldehyde was carried out, and a corresponding bis-azo group containing spirochromene was isolated. Using mass spectrum analysis, NMR, and IR spectroscopy, the structures of the produced compounds were identified.

KEYWORDS

- **bis-azo**
- **azo-compounds**
- **photochromic compound**
- **synthesis**
- **indolenine**

REFERENCES

1. Bukia, T.; Utiashvili, M.; Tsiskarishvili, M.; Jalalishvili, S.; Gogolashvili, A.; Tatrishvili, T.; Petriashvili, G. Synthesis of Some Azo Dyes on the Base of 2,3,3-Trimethyl-3*H*-Indolenine. *Chem. Chem. Techn.* **2023,** *17* (3).
2. Bukia, T.; Davitashvili, K.; Tsirekidze, N.; Markhulia, J.; Tatrishvili, T.; Akhobadze, S.; Petriashvili, G. Synthesis of a New Hybrid (Azo and Spiro) Photochromic Compound on the Base of 2,3,3-trimethyl-5-nitro-3*H*-indolenine. *Oxid. Commun.* **2023,** *46* (1), 177–187.
3. Petriahsvili, G.; Devadze, L.; Zurabishvili, T.; Sepashvili, N.; Chubinidze, K.; Tatrishvili, T.; Bukia, T.; Kalandia, E.; Areshidze, M.; Sharashidze, L.; Akhobadze, S.; Petriashvili, A.; Chubinidze, M.; Kinkladze, V.; Imnaishvili, N. Spiropyran Doped Liquid Crystal Film for the Monitoring of Human Ecology upon Exposure to UV Radiation: An Environmentally Friendly Approach. *Environ. Technol. Sustain.* **2022**.
4. Petriashvili G.; Devadze L.; Zurabishvili T.; Sepashvili N.; Chirakadze A.; Bukia T.; Markhulia J.; Areshidze, M.; Sharashidze, L.; Akhobadze, S.; Arveladze, E.; Sanikidze, G. Gold Nanoparticles Mediated Tuning of Thermo-Optical Parameters in Gold Nanoparticles Doped Cholesteric Liquid Crystal Nanocomposite. *Nano Studies* **2022,** 21–22.
5. Benkhaya, S.; Harfi, A. E.; Elharfi, A.; Classifications, Properties and Applications of Textile Dyes: A Review. *Appl. J. Envir. Eng. Sci.* **2017,** *3*(3), 311–320.
6. Mikroyannidis, J. A.; Tsagkournos, D. V.; Balraju, P.; Sharma G. D. Low Band Gap Dyes Based 3 on 2-styryl-5-phenylazo-pyrrole: Synthesis and Application for Efficient Dyesensitized Solar Cells. *J. Power Sources*. **2011,** *196,* 4152–4161. https://doi.org/10.1016/j.jpowsour.2010.12.038
7. Al-Ahmad, A. Y.; Hassan, Q. M. A.; Badran, H. A.; Hussain, K. A. Investigating Some Linear and Nonlinear Optical Properties of the Azo Dye (1-amino-2-hydroxy Naphthalin Sulfonic Acid-[3-7(4-azo)]-4-amino Diphenyl Sulfone). *Opt. Laser Technol.* **2012,** *44,* 1450–1455. https://doi.org/10.1016/j.optlastec.2011.12.019
8. Pevzner, E.; Ehrenberg, B. Principal Component Analysis of the Absorption and Resonance Raman Spectra of the Metallochromic Indicator Antipyrylazo III. *Spectrochim. Acta A Mol. Biomol.* **2000,** 56, 637–651. https://doi.org/10.1016/S1386-1425 (99)00164-X
9. Feng, Y.; Liu, H.; Luo, W.; Liu, E.; Zhao, N.; Yoshino, K.; Feng, W. Covalent Functionalization of Graphene by Azobenzene with Molecular Hydrogen Bonds for Long-Term Solar Thermal Storage. *Sci. Rep.* **2013,** *3* (1), 3260, 1–8. https://doi.org/10.1038/srep03260
10. Adegoke, O. A.; Adesuji, T. E.; Thomas, O. E. Novel Colorimetric Sensors for Cyanide Based on Azo-Hydrazone Tautomeric Skeletons. *Spectrochim. Acta A Mol. Biomol.* **2014,** *128,* 147–152. https://doi.org/10.1016/j.saa.2014.02.118
11. Coelho, P. J.; Carvalho, L. M.; Fonseca, A. M. C.; Raposo, M. M. Photochromic Properties of Thienylpyrrole Azo in Solution. *Tetrahedron. Lett.* **2006,** *47,* 3711–3714. https://doi.org/10.1016/j.tetlet.2006.03.125

12. Ruyffelaere, F.; Nardello, V.; Schmidt, R.; Aubry, J. M. Photosensitizing Properties and Reactivity of Aryl Azo Naphtol Dyes Towards Singlet Oxygen. *J. Photochem. Photobiol.* **2006,** *183,* 98–105. https://doi.org/10.1016/j.jphotochem.2006.02.029
13. Erfantalab, M.; Khanmohammadi, H. New 1.2.4-triazole-based Azo–Azomethine Dye. Part III: Synthesis, Characterization, Thermal Property, Spectrophotometric and Computational Studies. *Spectrochim. Acta A: Mol. Biomol.* **2014,** *125,* 345–352. https://doi.org/10.1016/j.saa.2014.01.113
14. Asgari-sabet, R.; Khoshsima, H. Real-time Holographic Investigation of Azo Dye Diffusion in a Nematic Liquid Crystal Host. *Dyes Pigm.* **2010,** *87,* 95–99. https://doi.org/10.1016/j.dyepig.2010.02.009
15. Kim, Y. D.; Cho, J. H.; Park, C. R.; Choi, J. H.; Yoon, C.; Kim, J. P. Synthesis, Application and Investigation of Structure–Thermal Stability Relationships of Thermally Stable Watersoluble Azo Naphthalene Dyes for LCD Red Color Filters. *Dyes Pigm.* [**2011,** *89,* 1–8. https://doi.org/10.1016/j.dyepig.2010.07.008
16. Avella-Oliver, M.; Morais, S.; Puchades, R.; Maquieira, Á. Towards Photochromic and Thermochromic Biosensing. *TrAC - Trends Anal. Chem.* **2016,** *79,* 37–45. https://doi.org/10.1016/j.trac.2015.11.021
17. Zhang, X.; Hou, L.; Samorı, P. Coupling Carbon Nanomaterials with Photochromic Molecules for the Generation of Optically Responsive Materials. *Nat. Commun.* **2016,** *7,* 11118. https://doi.org/10.1038/ncomms11118
18. Benkhaya, S.; M'rabet, S.; El Harfi, A. Classifications, Properties, Recent Synthesis and Applications of Azo Dyes. *Heliyon.* **2020,** *6,* https://doi.org/10.1016/j.heliyon.2020.e03271
19. Esser-Kahn, A. P.; Odom, S. A.; Sottos, N. R.; White, S. R.; Moore, J. S. Triggered Release from Polymer Capsules. *Macromolecules.* **2011,** *44,* 5539–5553. https://doi.org/10.1021/ma201014n
20. Yousefi, H.; Yahyazadeh, A.; Moradi Rufchahi, E.O.; Rassa, M. Synthesis, Spectral Properties, Biological Activity and Application of New 4-(benzyloxy)phenol Derived Azo Dyes for Polyester Fiber Dyeing. *J. Mol. Liq.* **2013,** *180,* 51–58. https://doi.org/10.1016/j.molliq.2012.12.030
21. Mallikarjuna, N. M.; Keshavayya, J. Synthesis, Spectroscopic Characterization and Pharmacological Studies on Novel Sulfamethaxazole based Azo Dyes. *J. King. Saud. Univ. Sci.* **2009,** *32,* 251–259.
22. Yazdanbakhsh, M. R.; Yousefi, H.; Mamaghani, M.; Moradi, E.O.; Rassa, M.; Pouramir, H.; Bagheri, M. Synthesis, Spectral Characterization and Antimicrobial Activity of Some New Azo 3 Dyes Derived from 4.6 Dihydroxypyrimidine, *J. Mol. Liq.* **2012,** *169,* 21–26. https://doi.org/10.1016/j.molliq.2012.03.003
23. Karcı, F.; Şener, N.; Yamac, M.; Şener, İ.; Demircalı, A. A. The Synthesis, Antimicrobial Activity and Absorption Characteristics of Some Novel Heterocyclic Disazo Dyes. *Dyes Pigments.* **2009,** *80,* 8047–8052. https://doi.org/10.1016/j.dyepig.2008.05.001.
24. Xu, H.; Zeng, X. Synthesis of Diaryl-Azo Derivatives as Potential Antifungal Agents. *Bioorgan Med. Chem. Lett.* **2010,** *20,* 4193–4195. https://doi.org/10.1016/j.bmcl.2010.05.048

25. Liu, G.; Wang, X.; Hu, J.; Zhang, G.; Liu, S. Self-Immolative Polymersomes for High-Efficiency Triggered Release and Programmed Enzymatic Reactions. *J. Am. Chem. Soc.* **2014**, *136,* 7492–7497. https://doi.org/10.1021/ja5030832
26. Hunger, K.; Mischke, P. Azo Dyes, 1. *Ullmann's Encycl. Ind. Chem.* **2011,** *4,* 523–541. https://doi.org/10.1002/14356007.a03_245.pub2
27. Hunger, K.; Mischke, P.; Rieper, W.; Raue, R.; Kunde, K.; Aloys E. Azo Dyes. *Ullmann's Encycl. Ind. Chem.* **2000,** 1–93. https://doi.org/10.1002/14356007.a03_245
28. Rydchuk, M.; Vrublevska T.; Boyko M.; Korkuna, O. Masking is the Effective Alternative to the Separation during Osmium Determination by Means of Azo Dyes in Complex Samples. *Chem. Chem. Technol.* **2010**, *4* (2), 115–124. https://doi.org/10.23939/chcht04.02.115.
29. Kulkarni, M.; Thakur, P. The Effect of Uv/Tio2 /H2o2 Process and Influence of Operational Parameters on Photocatalytic Degradation of Azo Dye in Aqueous Tio2 Suspension. *Chem. Chem. Technol.* **2010**. *4* (4), 265–270. https://doi.org/10.23939/chcht04.04.265
30. Cardanol, F.; Frasconi, M.; Giordani S. Photo-Responsive Graphene and Carbon Nanotubes to Control and Tackle Biological Systems. *Front. Chem.* **2018,** *6,* 12. https://doi.org/10.3389/fchem.2018.00102

CHAPTER 4

Comb-Type Organosilicon Polymers with Photoswitchable Fragments

TAMARA TATRISHVILI[1,2] and KALOIAN KOYNOV[3]

[1]*Department of Macromolecular Chemistry, Ivane Javakhishvili' Tbilisi State University, Tbilisi, Georgia*

[2]*Institute of Macromolecular Chemistry and Polymeric Materials, Ivane Javakhishvili Tbilisi State University, Tbilisi, Georgia*

[3]*Max Planck Institute for Polymer Research, Ackermannweg, Mainz, Germany*

ABSTRACT

The aim of this work is a synthesis of polymers with photoswitchable fragments. As the first step, we synthesized the initial compounds, 4-((4-fluorophenyl) diazenyl)phenol and 1-(4-allyloxy)phenyl)-2-(4-fluorophenyl) diazene. Hydrosilylation reaction of 2.4.6.8-tetrahydro-2.4.6.8-tetramethylcyclotetrasiloxane with 4-((4-(allyloxy)phenyl)diazenyl)phenylpropionate as well as with vinyltriethoxysilane was conducted.

Polymerization and the co-polymerization reaction of organocyclotetrasiloxanes with terminated agent hexamethyldisiloxane in the presence of a nucleophilic catalyst were carried out; Investigation of sol-gel reactions of ethoxy-containing compounds and polymers for obtaining these films was performed.

Also, nuclear magnetic resonance (NMR) spectroscopy, differential scanning calorimetry (DSC), thermogravimetric analysis (TGA), and gel

Advances in Engineering Materials: Preparation for Sustainable Process Development. Tamara Tatrishvili and A. K. Haghi (Eds.)

permeation chromatography (GPC) analysis of obtaining polymers and polymeric materials were fulfilled.

4.1 INTRODUCTION

Azobenzene, consisting of two phenyl rings separated by an azo (–N=N–) bond, acts as the precursor to a wide range of aromatic azo compounds. The photochromic trans-cis isomerization of azobenzenes has been extensively applied in molecular switches, utilizing its shape alteration upon isomerization to generate a supramolecular outcome. Azobenzenes incorporated into the polymer backbone provide switchable receptors, while those in monolayers yield surface properties that can be light-controlled.

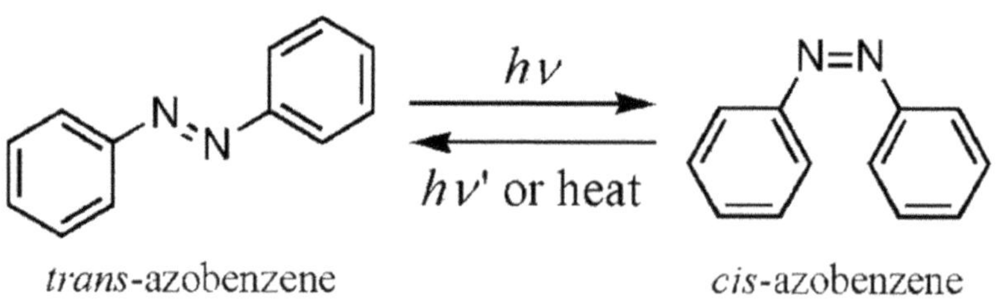

These chromophores are adaptable molecules and have garnered considerable interest in fundamental and applied research domains. The electronic absorption band can be customized by ring substitution, resulting in a spectrum ranging from ultraviolet to red visible regions, thereby allowing the precise adjustment of color.

This versatility, coupled with the inherent chemical stability and durability of the azo groups, has prompted extensive investigations into azobenzene-based structures as dyes and colorants.

One of the fascinating features of these chromophores is the easily induced and reversible trans-cis photoisomerization of the azo bond and the ensuing geometric alterations when azo chromophores are integrated into various materials, including polymers.[1,2]

Azobenzene-based photocontrol has been demonstrated in a range of applications, including photoswitchable phase changes,[3] phase separation (or reversal of phase separation),[4,5] solubility changes,[6,7] and crystallization.[8,9]

This research aims to synthesize new comb-type organosilicon polymers that incorporate azobenzene side groups within the polymer chain. Inserting azobenzene chromophore groups into the siloxane matrix enhances the flexibility, hydrophobicity, and thermal stability of the resulting polymers and films compared to their organic counterparts.

4.2 EXPERIMENTAL METHODS AND MATERIALS

4.2.1 GENERAL CHARACTERIZATION

The obtained monomers, polymers, and polymer films have been carefully characterized at all stages of preparation using various experimental techniques.

FT-IR and NMR spectroscopy

FT-IR and NMR spectroscopy were used to study the synthesized materials at different stages of the synthesis and confirm their chemical structures and compositions.

GPC and DSC

Gel permeation chromatography (GPC) was used to estimate the molecular weight of the prepared polymers. Differential scanning calorimetry (DSC) will be used to study the temperature characteristics of the polymers. In particular, we planned to study the correlation between the polymer structure and the glass transition, crystallization, and melting temperatures.

Dynamic mechanical analysis

We have investigated the viscoelastic properties of the prepared polymer films. Temperature and frequency dependencies of the real (G′) and imaginary (G″) parts of the complex shear modulus have been measured. Such studies revealed important information about the mechanical properties of these systems, the glass transition temperature, and eventual melting and crystallization processes. These findings were compared with the DSC studies to achieve a more complete understanding.

Tensile tests

These studies reveal important information about the static elastic properties of the films. Stress vs. draw ratio dependencies were recorded and used to evaluate the elastic modulus, elongation at break, and stress at break of the samples. Furthermore, the performance of the materials will be studied at large deformations by performing quasi-static stress-strain cycles with successively increasing maximum strain. These studies may reveal important information for plastic deformation resulting from changes in the internal structure of the materials.

Photophysical Characterization

The photochromic switching from the *trans*- to the *cis*- form (UV light) and from the *cis*- to the *trans*- form (VIS) in solutions of the polymers was systematically investigated. The results have been compared with similar studies for solutions of the pure azo dyes to evaluate the influence of the covalent attachment to the polymer chain on the photochromic properties.

As a next step, thin films on a substrate or stay-alone films from the polymers were prepared, and their photochromic properties were studied. The photochromic properties of the films were evaluated and compared with the solution results. Finally, the effects of the photoisomerization on the tensile mechanical properties of the films were studied by additional stress-strain experiments.

4.3 RESULTS AND DISCUSSION

4.3.1 SYNTHESIS OF MONOMERS

For obtaining 1-(4-allyloxy) phenyl)-2-(4-fluorophenyl) diazene in the first stage, the diazotization (azo combination) reaction of 4-fluoroaniline with nitric acid was carried out, and then the combination of obtained diazonium salt with phenol was carried out and corresponding 4-((4-fluorophenyl)diazenyl)phenol will be obtained according to the following Scheme 4.1:

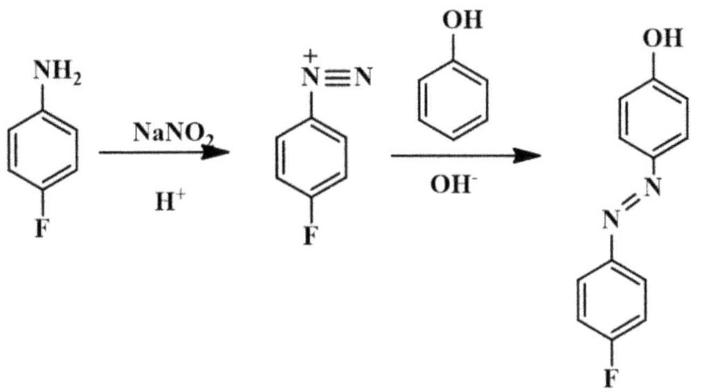

(I)

SCHEME 4.1 Azo combination reaction of 4-fluoroaniline and obtaining of 4-((4-fluorophenyl) diazenyl) phenol.

The structure and composition of obtained compound I was determined via FT-IR, ^{1}H, ^{13}C, ^{29}Si NMR spectra data.

In ^{1}H NMR spectra of the compound, I observed multiple signals for aryl fragments with chemical shifts in the range $\delta = 7–8$ ppm, as well as singlet signals for aromatic-OH groups with chemical shift $\delta = 3.4$ ppm.

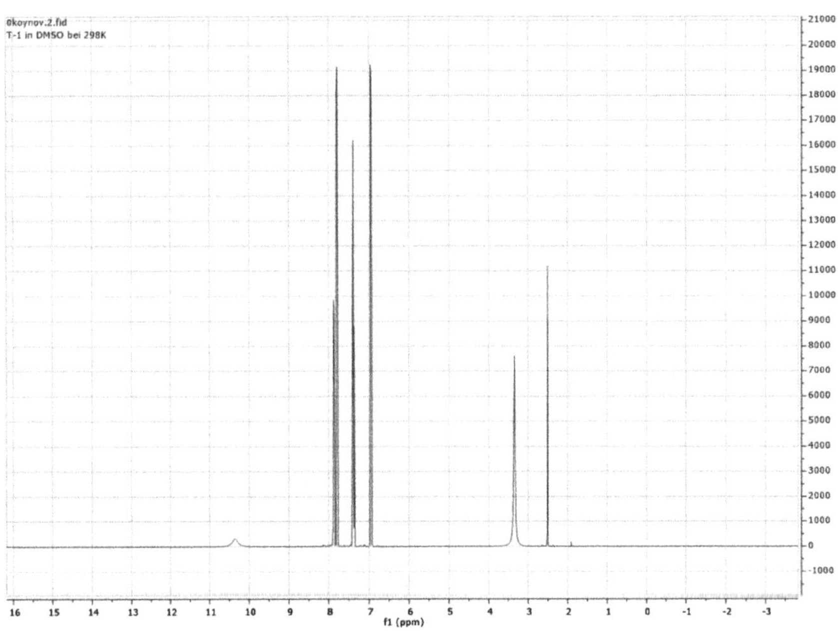

FIGURE 4.1 ^{1}H NMR spectra of compound I.

In ^{13}C NMR spectra of compound I, the values of the shifts of the carbon atoms are presented on the compound I structure.

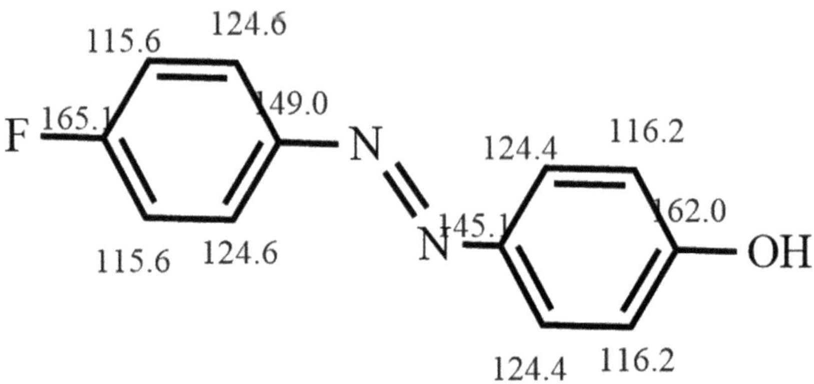

In the second stage, the reaction of 4-((4-fluorophenyl) diazenyl) phenol with allyl bromide will be carried out, and 1-(4-allyloxy) phenyl)-2-(4-fluorophenyl) diazene will be obtained according to the Scheme 4.2.

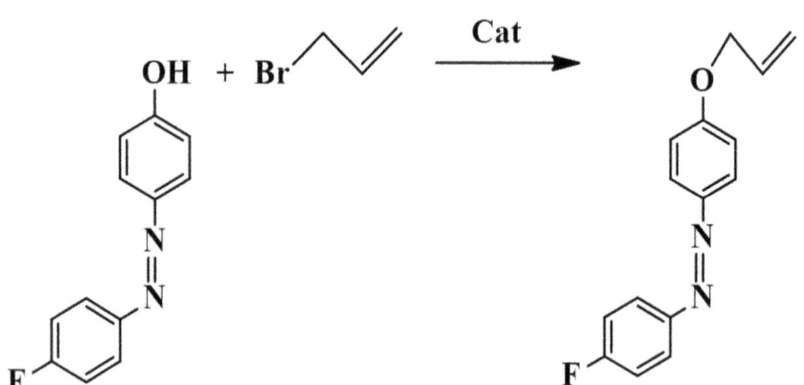

(II)

SCHEME 4.2 Condensation reaction 4-((4-fluorophenyl) diazenyl) phenol with allyl bromide.

The structure and composition of the obtained compounds were studied via FT-IR, ^{1}H, ^{13}C NMR spectra data, and determination of molecular masses.

In ^{1}H NMR spectra of compound II, one can observe multiple signals for aryl fragments with chemical shifts in the range $\delta = 6.7–7.8$ ppm, as well as multiple signals for methylene protons in O-CH_2- groups with chemical shift $\delta = 4.7$ ppm, multiple signals for protons in =CH_2 group with chemical shifts at $\delta = 5.3$ and 5.5 ppm.

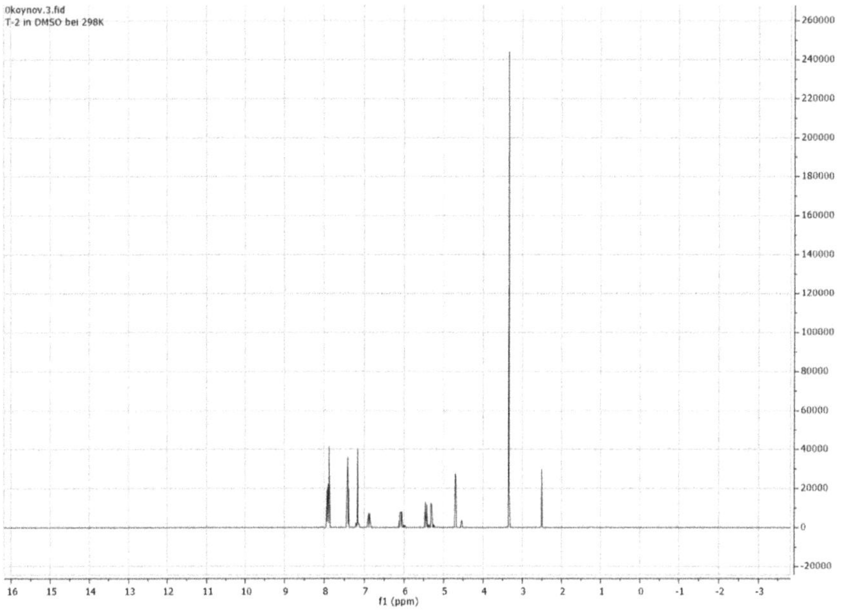

FIGURE 4.2 ^{1}H NMR spectra of compound II.

In ^{13}C NMR spectra of compound II, the chemical shifts of the groups are presented on the structure of compound II:

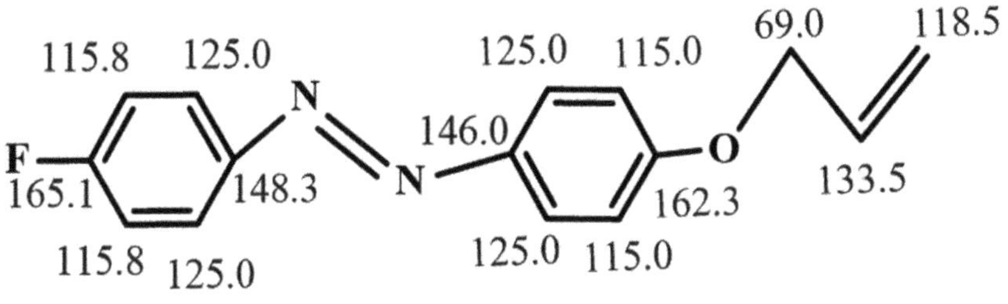

The hydrosilylation reaction of 2.4.6.8-tetrahydro-2.4.6.8-octamethyl-cyclotetrasiloxane with 4-((4-(allyloxy) phenyl) diazenyl) phenylpropionate as well as with vinyltriethoxysilane was performed in the presence of a platinum catalyst and corresponding adducts was obtained, according to the following Schemes 4.3 and 4.4:

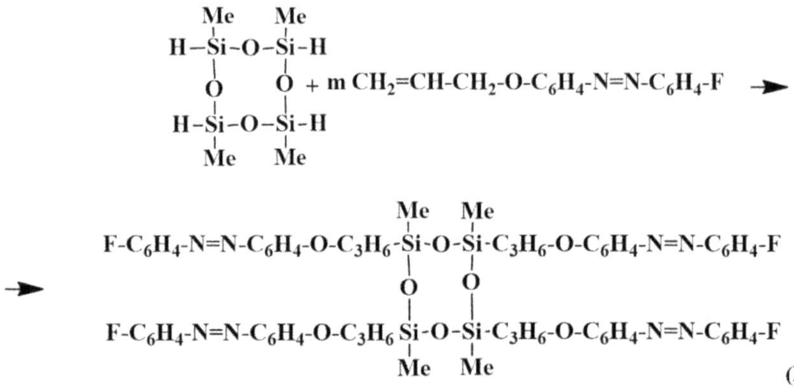

(III)

SCHEME 4.3 Hydrosilylation reactions of 2.4.6.8-tetrahydro-2.4.6.8-tetramethylcyclotetrasiloxane with 1-(4-allyloxy) phenyl)-2-(4-fluorophenyl) diazene (ratio 1:4).

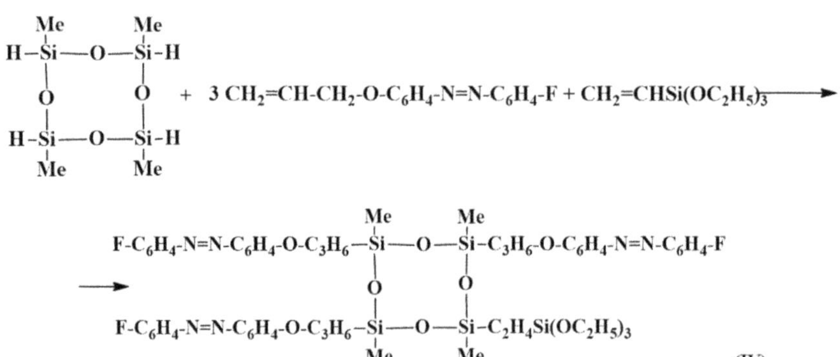

(IV)

SCHEME 4.4 Hydrosilylation reactions of 2.4.6.8-tetrahydro-2.4.6.8-tetramethylcyclotetrasiloxane with 1-(4-allyloxy)phenyl)-2-(4-fluorophenyl)diazene and vinyltriethoxysilane (ratio 1:3:1).

The structure and composition of obtained compounds II–IV have been determined via FT-IR, ^{1}H, ^{13}C, ^{29}Si NMR spectra, and molecular masses.

In the spectra of compound IV, additionally, one can observe the triplet signals with chemical shift $\delta = 0.6$ ppm, 1.0 ppm, and 1.8 ppm multiple signals with chemical shift $\delta = 3.8$ ppm characteristic for methylene and methyl protons in $CH_3CH_2OSiCH_2$, $CH_2CH_2CH_2SiMe$, CH_3CH_2OSi, and CH_3CH_2OSi accordingly.

^{13}C NMR spectra of compound IV are in accordance with ^1H NMR spectra data.

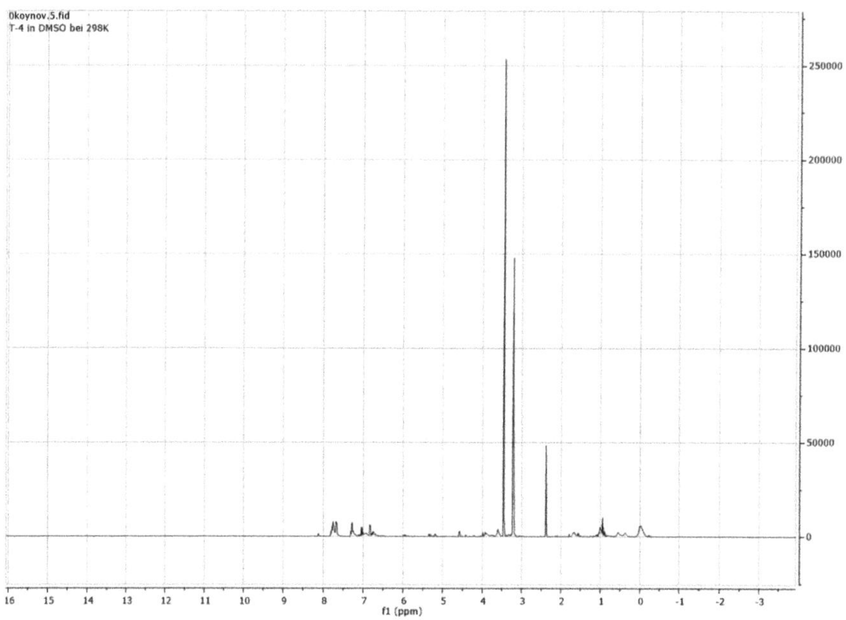

FIGURE 4.3 ^1H NMR spectra of compound IV.

4.3.2 *SYNTHESIS OF POLYMERS*

The polymerization reaction of compounds III and IV has been carried out in the presence of nucleophilic catalyst potassium hydroxide at different temperatures and quantities of catalyst in the presence of terminating agent hexamethyldisiloxane or without it, according to Schemes 4.5 and 4.6:

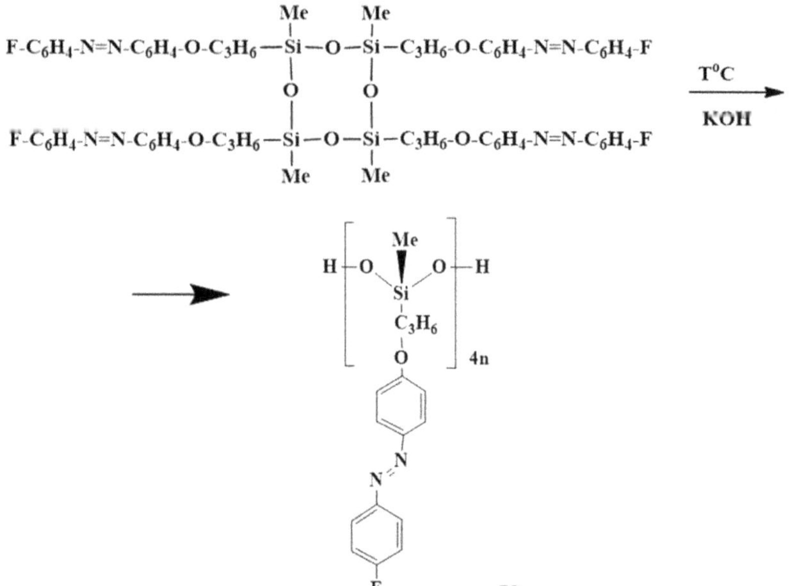

(V)

SCHEME 4.5 Polymerization reaction of compound III.

The structure and composition of oligomer V have been determined via FT-IR, ^{1}H, ^{13}C, ^{29}Si NMR spectra data. The chemical shifts of the groups are presented on the structure.

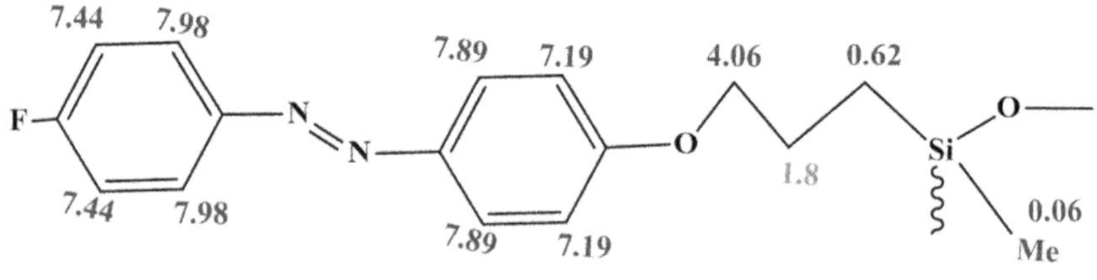

For synthesized oligomer V GPC investigation has been carried out.

Likewise, for obtained oligomer V Differential Scanning Calorimetry, an investigation was carried out, and the glass transition temperature was determined. DSC investigation has been provided on Mettler Toledo DSC-822.

Via differential scanning calorimetric investigation, it was observed three endothermic peaks in the region 100–150°C, which may be corresponds to cis and trans forms of azobenzene fragments.

Molecular weight distribution:

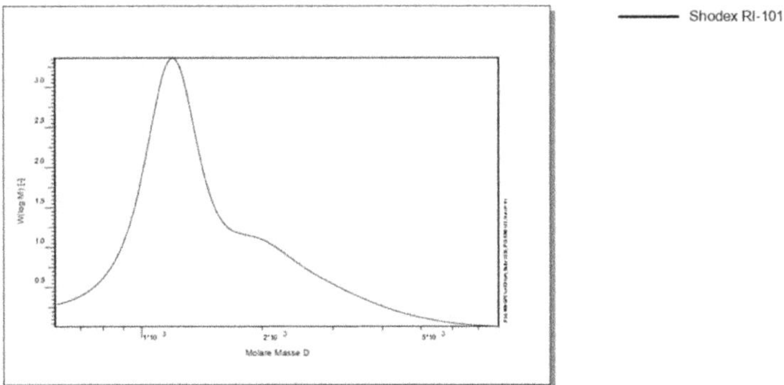

Name	Mn	Mw	D	Vp	Mp	Area
Shodex RI-10	1340,84	1627,01	1,21	30,73	1198,63	0,40

FIGURE 4.4 Gel permeation chromatographic curve of compound V.

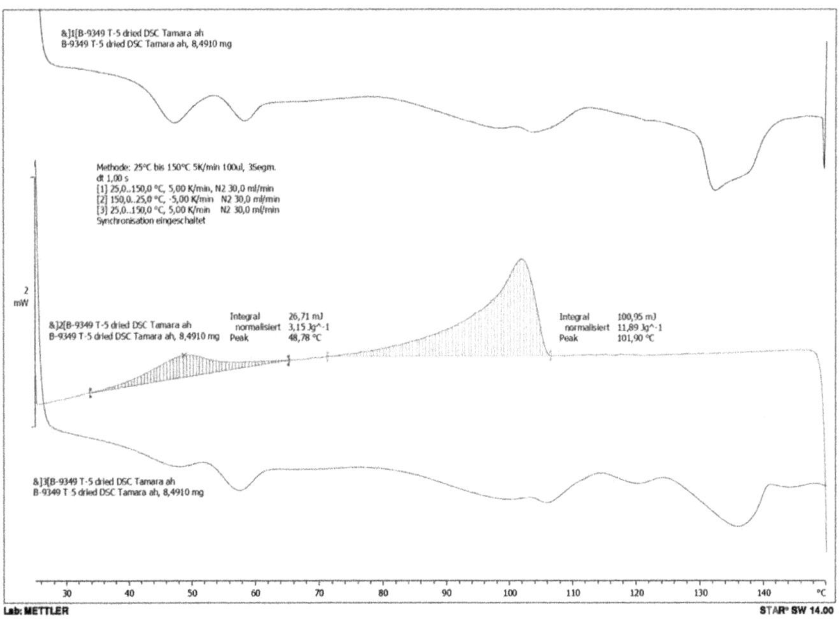

FIGURE 4.5 DSC investigation of oligomer V.

TGA investigations have been carried out on the Mettler Toledo DSC-822 device. The heating rate was 10°C/min.

In Figure 4.6, the thermogravimetric curve of oligomer V is presented.

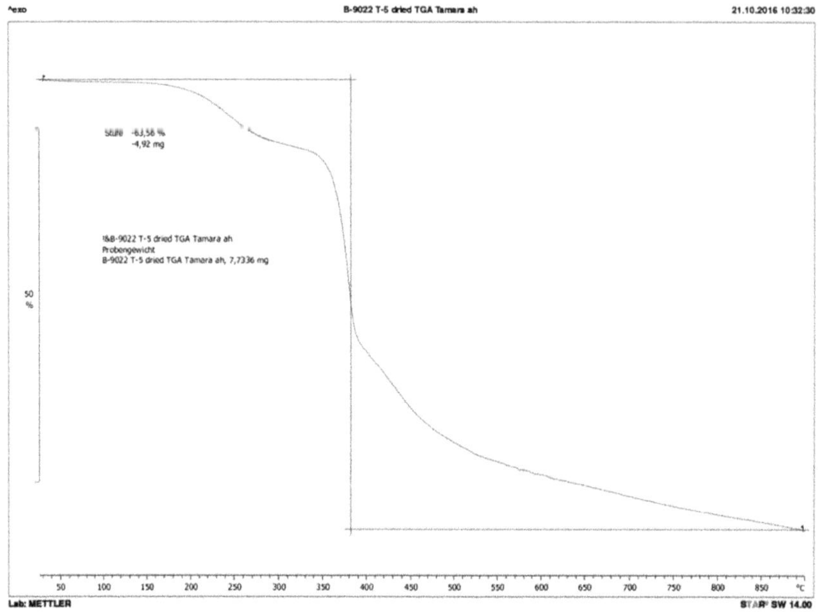

FIGURE 4.6 TGA of oligomer V.

As it is seen from Figure 4.6, 4–5% mass losses are observed at 200°C. The primary destruction process proceeds in the temperature range of 350–500°C. The residual mass of the oligomer at 850°C is equal to 63.5%.

The polymerization reaction of the compound was conducted according to the scheme:

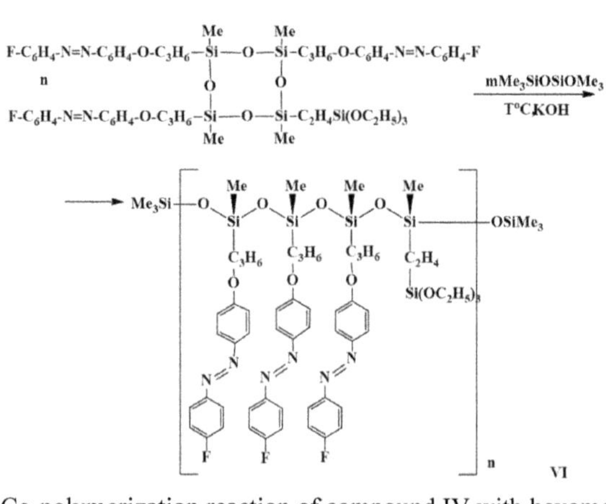

VI

SCHEME 4.6 Co-polymerization reaction of compound IV with hexamethyldisiloxane.

The obtained polymers were studied by FT-IR, ^{1}H, ^{13}C, ^{29}Si NMR spectra, GPC and DSC analyses, which confirmed the structure of the polymers.

4.3.3 PREPARATION OF POLYMER FILMS

Film preparation via sol–gel processes of compound IV and polymer VI were carried out, and thin films were obtained according to Schemes 4.7 and 4.8:

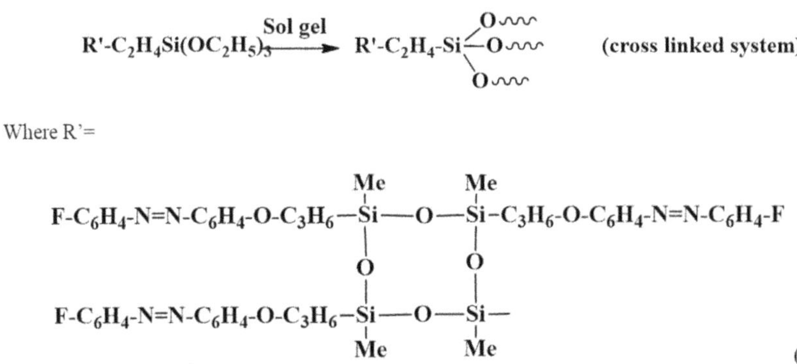

(VII)

SCHEME 4.7 Sol–gel reaction of compound IV.

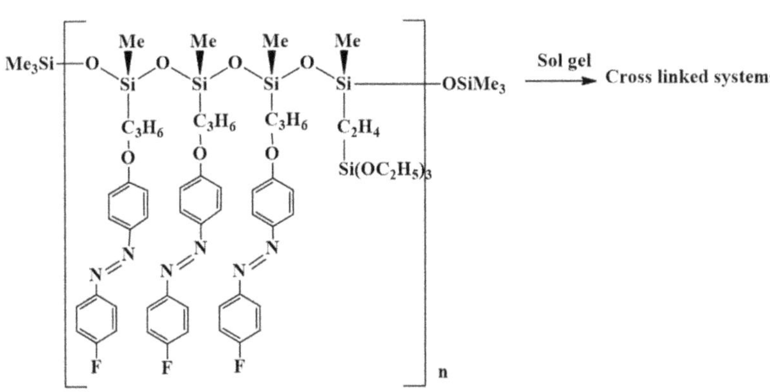

(VIII)

SCHEME 4.8 Sol–gel reaction of polymer VI.

As shown in Figures 4.7 and 4.8, the UV investigations of compound II and oligomer V have been carried out. The *-trans* form (on the left) can be transformed into the *-cis* form (on the right) through exposure to an appropriate wavelength (UV at 300–400 nm) of light.

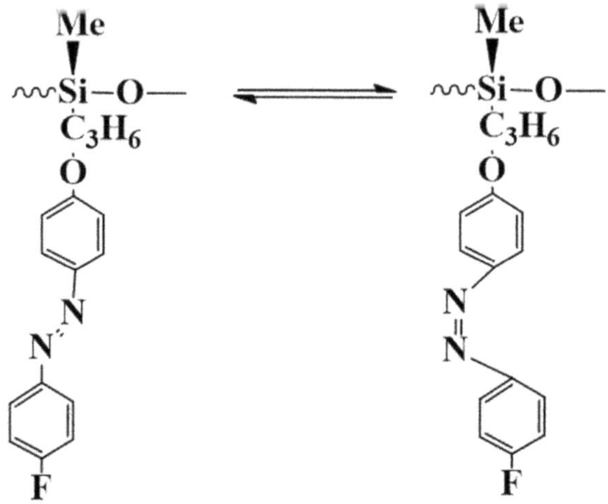

To convert the molecule back to the *-trans* form, a distinct wavelength (visible blue light >400 nm) can be utilized. Alternatively, the molecule will return to the stable trans form through thermal relaxation.

Sol–gel reactions of compound IV and oligomer V have been carried out, but unfortunately, rigid films have been obtained. So, we cannot study the mechanical properties of these films.

4.4 CONCLUSION

For obtaining and characterization of organosilicon polymers with photo-switchable fragments, first of all where synthesized initial monomers: 4-((4-fluorophenyl)diazenyl)phenol, which was used in the next step of synthesis and 1-(4-allyloxy)phenyl)-2-(4-fluorophenyl)diazene. The structure and composition of monomers were confirmed via FT-IR and NMR spectral analysis.

By hydrosilylation reaction of 2.4.6.8-tetrahydro-2.4.6.8-tetramethylcyclotetrasiloxane with 4-((4-(allyloxy)phenyl)diazenyl)phenyl propionate as well as with vinyltriethoxysilane were performed. There were synthesized cycling monomers with fluorophenyl-diazene and vinytrhiethoxysilyl fragments in the side chain.

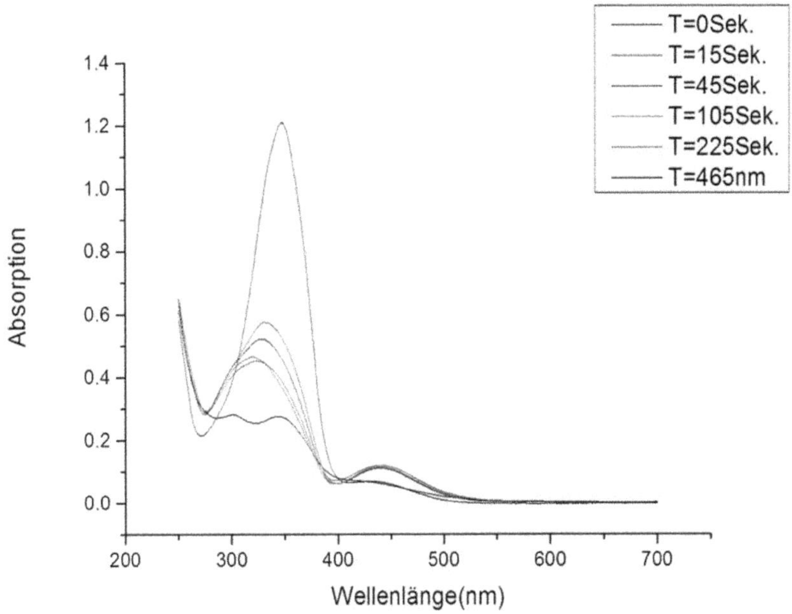

FIGURE 4.7 Photoisomerization of compound II.

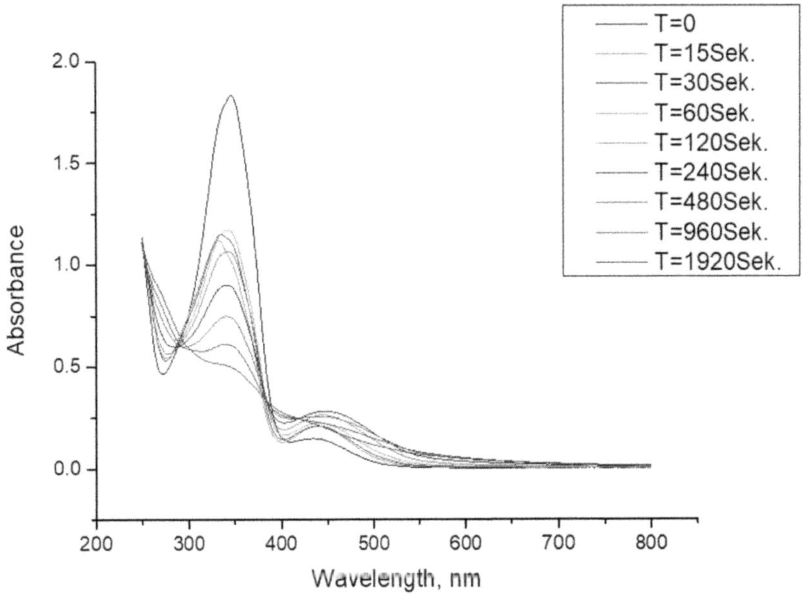

FIGURE 4.8 Photoisomerization of oligomer V.

Based on obtained monomers, polymerization reactions have been carried out in the presence of nucleophilic potassium hydroxide, at various temperatures and amounts of catalyst in the presence of terminating agent hexamethyldisiloxane or without it.

Combe-type methyl siloxane polymers with photoswitchable fragments in the side chain were performed. The obtained polymers were studied by FT-IR, ^{1}H, ^{13}C, ^{29}Si NMR spectra, GPC, and DSC analyses.

Via sol–gel processes of polymers film preparation have been carried out, but unfortunately, rigid films have been obtained.

KEYWORDS

- **photoswitchable**
- **hydrosilylation reaction**
- **FT-IR**
- **NMR spectra data**
- **sol-gel**
- **UV spectroscopy**

REFERENCES

1. Delaire, J. A.; Nakatani, K. Linear and Nonlinear Optical Properties of Photochromic Molecules and Materials. Chem. Rev. **2000,** *100* (5), 1817–1846.
2. Yesodha, S. K.; Sadashiva Pillai, C. K.; Tsutsumi, N. Stable Polymeric Materials for Nonlinear Optics: A Review Based on Azobenzene Systems. *Prog. Polym. Sci.* **2004,** *29* (1), 45–74.
3. Aoki, K.; Nakagawa, M.; Ichimura, K. Self-assembly of Amphoteric Azopyridine Carboxylic Acids: Organized Structures and Macroscopic Organized Morphology Influenced by Heat, ph Change, and Light. *J. Am. Chem. Soc.* **2000,** *122* (44), 10997–11004.
4. Kadota, S.; Aoki, K.; Nagano, S.; Seki, T. Photocontrolled Microphase Separation of Block Copolymers in Two Dimensions. *J. Am. Chem. Soc.* **2005,** *127* (23):8266–8267.
5. Effing, J. J.; Kwak, J. C. T. Photoswitchable Phase Separation in Hydrophobically Modified Polyacrylamide/Surfactant Systems. *Angew. Chem. Int. Ed.* **1995,** *34* (1):88–90.

6. Yamamoto, H.; Nishida, A.; Takimoto, T.; Nagai, A. Photoresponsive Peptide and Polypeptide Systems. VIII. Synthesis and Reversible Photochromism of Azo Aromatic Poly(L-ornithine). *J. Polym. Sci. Part A Pol. Chem.* **1990,** *28* (1), 67–74.
7. Arai, K.; Kawabata, Y. Changes in the Sol-Gel Transformation Behavior of Azobenzene Moietycontaining Methyl Cellulose Irradiated with UV Light. *Macromol. Rapid. Commun.* **1995,** *16* (12), 875–880.
8. Ebralidze, T. D.; Mumladze, A. N. Light-induced Anisotropy in Azo-Dye-Colored Materials. *Appl. Opt.* **1990,** *29* (4), 446–447.
9. Tarasenko, V.; Nadtoka, O.; Syromyatnikov, V. Side Chains Azobenzene Moieties in Polymethacrylates for lc Alignment. In *High-Performance Polymers for Engineering-Based Composites*; Mukbaniani, O. V., Abadie, M. J. M.; Tatrishvili, T. Eds.; Apple Academic Press, Inc., 2015; p 325.

CHAPTER 5

Polyamines and Arginine-Based Cationic Polymers as Antimicrobial Agents

NINO ZAVRADASHVILI[1], TAIA KOTORASHVILI[1], TENGIZ KANTARIA[1], NINO KUPATADZE[1], NATIA OCHKHIKIDZE[1,2], NINO MEDZMARIASHVILI[2], NAZI KUTSIAVA[2], MANANA GURIELIDZE[3], and RAMAZ KATSARAVA[1,2]

[1]*Institute of Chemistry and Molecular Engineering, Agricultural University of Georgia, Tbilisi, Georgia*

[2]*Department of Chemical and Biological Technologies, Georgian Technical University, Tbilisi, Georgia*

[3]*Durmishidze Institute of Biochemistry and Biotechnology, Agricultural University of Georgia, Tbilisi, Georgia*

ABSTRACT

Prevention and treatment of infectious diseases, especially those revealing antibiotic resistance, is a topical problem today. Among various synthetic antibacterial agents, cationic polymers (CPs) are becoming increasingly promising materials. CPs are also of interest as gene intracellular delivery (transfection) agents for gene therapy and biotechnology. Although, the main limitations of the current CPs (such as polyethylenimine [PEI], poly-L-lysine [PLL], poly-L-arginine [poly-Arg], or polyamidoamine-based dendrimers, etc.) as either antibacterials or gene transfection agents are a high cytotoxicity and/or stability (nonbiodegradability) in the physiological conditions. We have developed new biodegradable CPs composed of polyamines—endogenous tetraamine spermine (Spm) and

Advances in Engineering Materials: Preparation for Sustainable Process Development. Tamara Tatrishvili and A. K. Haghi (Eds.)

related triamine with shorter polymethylene chains—*N*-(2-aminoethyl)-1,3-propanediamine (Apd) for incorporating *sec*-amino groups to the polymeric backbones. The highly-charged guanidine groups in the goal CPs were incorporated using the naturally occurring amino acid arginine (R). Some representatives of water-soluble CPs revealed rather high antimicrobial activity (tested against *Escherichia coli*, *Salmonella enterica*, *Staphylococcus aureus*, *Bacillus cereus*). Hence, synthesized CPs could be interesting as they are low-toxic and potentially biodegradable antimicrobial agents.

5.1 INTRODUCTION

Cationic polymers (CPs) have emerged as a promising alternative to viral vectors in gene therapy and nanotechnology due to their gene transfection function. In addition, inherent bioactive properties such as antimicrobial, antioxidant, antitumor, and anti-inflammatory stimuli response make CPs more promising for enhanced therapeutic potential.[1,2] Biodegradable CPs, which can be cleared from the body after executing their function, are particularly valuable. Therefore, our research interest is focused on a systematic study of the design of new biodegradable CPs to be used as gene carriers and antimicrobial agents.

The use of CPs as bactericidal agents is gradually increasing.[3] In contrast to low-molecular-weight analogs, polymeric bactericides exhibit prolonged action and higher antimicrobial efficacy due to the high local concentration of active groups. The highly charged cationic molecule was found to disrupt the outer membrane and partially depolarize the inner membrane of bacteria.[4,5] Furthermore, it has been shown that cationic molecule is widely distributed in the bacterial cell at lethal concentrations and could bind to DNA. The cationic molecule's multimodal action reduces the likelihood of resistance development compared to single-target antibiotics.[5] This bactericidal mechanism may apply to other positively charged organic molecules, particularly highly charged CPs. For all the pharmacological applications, preference should be given to biodegradable CPs, which are removed from the body when their purpose has been fulfilled. Along with this, it is essential that the CPs are degraded, releasing nontoxic and physiological debris that will either be removed from the body without complications or assimilated by it via metabolic means. Accordingly, the building blocks used for constructing biodegradable CPs should be nontoxic and physiological. From this perspective, the positively charged naturally occurring molecules—polyamines and α-amino acids look especially promising.

A large number of linear polyamines have been detected in mammalian tissues and excreta or plants, bacteria, and microorganisms.[6] The molecules of polyamines start with a primary amino group (H_2N–) linked to a chain of methylene groups (–CH_2–) interspersed with secondary amino groups (–NH), and terminates with another primary amino group (–NH_2).[6] These molecules look suitable for constructing CPs having polyethylenimine (PEI)-like constitution—the terminal primary amino groups could be used for constructing macrochains via step-growth polymerization (polycondensation), and the secondary amino groups for incorporating PEI-like moieties in the polymeric backbones. Among a large number of polyamines, the most widely applicable for constructing CPs is an endogenous multiamino monomer (tetraamine), which professionally condenses genes in the body—spermine (Spm). Among the positively charged amino acids, L-arginine (R)—is a crucial component of many naturally occurring cationic peptides and proteins. It is of particular interest as a building block for constructing synthetic CPs.

The antimicrobial activity of Spm (as tetrachloride) and related compound—spermidine (as phosphate) was described in the early 1950s (see Ref. [7] and references cited therein). It was also shown that various derivatives of Spm possess significant antimicrobial activity. The susceptibility of *Escherichia coli* to hydrophobic antibiotics such as rifampicin was significantly increased by homologous Spm bis-acylated with C8 or C9 chains.[8] Spm, along with other polyamines, was used as a component of antimicrobial compositions.[9] *N*-Terminally modified dipeptidomimetics containing linear or branched Spm backbone,[10] as well as Spm alkaloids, also reveal a noticeable bactericidal activity.[11–13] Spm showed a strong synergistic antibacterial effect in combination with β-lactams and chloramphenicol.[14]

Amino acid arginine (R) attracts attention as a building block for antimicrobial compounds since it contains guanidine (iminourea) H_2N–(C=NH)–NH_2 group in the lateral chain—it is well known that a huge group of synthetic bactericides also contain guanidine group.[15–18] Less is known about the antimicrobial properties of pure amino acid arginine. However, lipoamino acid derived from arginine (N-acyl-arginine methyl ester, arginine-*N*-alkyl amide, and arginine-*O*-alkyl ester), which represents a family of nontoxic and biodegradable cationic surfactants, showed antimicrobial activity against a broad spectrum of bacteria.[19,20] Amino acid arginine is one of the most important components of antimicrobial peptides.[21–23] Despite this, there is very scant data in the literature regarding the antibacterial activity of polymers constituted of arginine. The most studied are chitosan derivatives with covalently attached R through free amino groups of the polysaccharide.[24–26] It was shown that poly-R/hyaluronan nanolayered film at a chain length of poly-R containing 30 residues revealed unexpected bactericidal activity.[27] It was shown recently that R-based biodegradable CPs such as poly(ester urea)s and poly(ester urethane)s reported in Ref. [28], revealed antimicrobial activity against different bacterial strains.[29] These polymers could be considered as high-molecular-weight analogs of lipoamino acid reported in.[19,20]

Two classes of polyamine and Spm-based CPs were synthesized (Scheme 5.1).[30] The resulting CPs—polyureas (PUs) and polyamides (PAs) such as polysuccinamides (PSuAs) showed high water solubility ($\leq$ 200 mg/mL)—which is very important for their practical applications. The CPs showed high in vitro cell compatibility within the concentration range of 0.04–10.00 mg/m (assessed using mammalian cell lines such as 4T1 (murine mammary carcinoma), human cervical adenocarcinoma (HeLa), NIH3T3 cells (mouse fibroblast) and CCL 210 (human fibroblast). The CPs formed positively charged (zeta potential $\leq$ +57.4 mV) stable nanocomplexes with the pDNA (38.0–312.0 nm) at a low weight ratio of polymer/pDNA (WR = 5/1–25/1). The cellular uptake of CP/pDNA (eGFP-encoding) complexes was evaluated using flow cytometry (fluorescence-activated cell sorting [FACS]) analysis and confocal microscopy study. Selected samples of Spm-based polymers showed relatively high and less known data selective transfection activity with various cell lines tested. They were considered as nontoxic and effective nanocarriers for their potential application in gene therapy.[30]

In this study, we have evaluated the antimicrobial activity of polyamines (Spm and Apd) and arginine (R) based CPs[1] (Scheme 5.1). The main characteristics of CPs synthesized are summarized in Table 5.1. A high dispersity of some polymers ($Đ$ > 2) can speak for the formation of branched polymers due to side reactions through sec-amino groups of Spm and Apd. According to the biological studies performed, the CPs have a

[1]All polymers synthesized were in salt forms—either *p*-toluenesulfonates (T) or hydrochlorides (C); in the case of RPAs, one hydrochloric acid molecule (C) belongs to the residue of R-methyl ester.

potential for use in gene therapy and biotechnology as nontoxic and effective nonviral gene delivery vectors.[30]

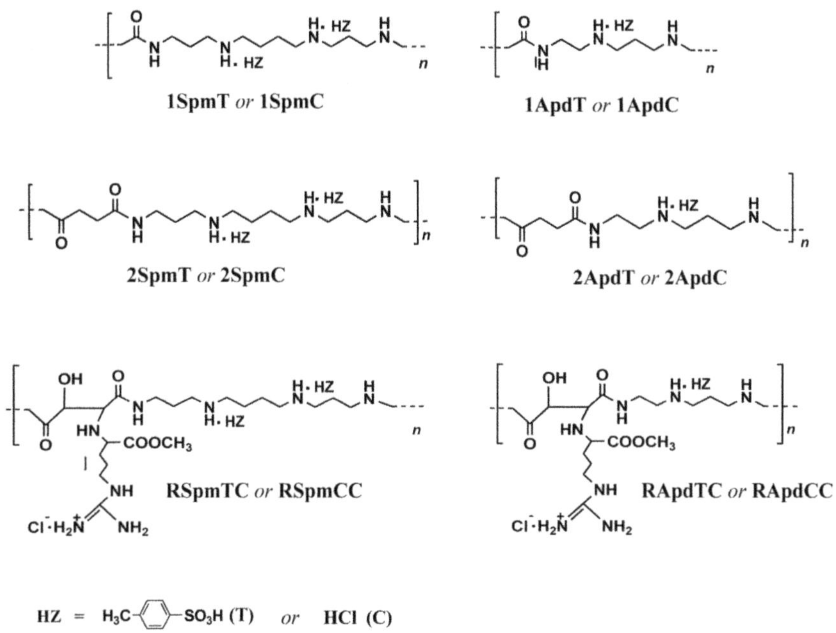

SCHEME 5.1 Structure of the Spm and Apd-based polymers: PAs, PUs and RPAs.

5.2 EXPERIMENTAL METHODS AND MATERIALS

In vitro antimicrobial activity study. Antimicrobial activity of new CPs against different strains of bacterial species: *Bacillus cereus*, *Staphylococcus aureus*, *E. coli*, *Salmonella enterica* was performed *via* the standard protocols of disk diffusion method (DDM) using corresponding American Type Culture Collection (ATCC) cultures and resistant strains from the stock by the presence-absence of inhibition zones values. Muller Hinton agar medium was prepared for the *in vitro* antibacterial activity study. The medium was poured into plates and remained to solidify. 0.1 mL of the bacterial cultures were inoculated through the spread plate technique. Whatman's filter paper discs (5 mm diameter) were soaked in 5% solutions of CPs. The soaked discs were placed on the surface of the inoculated plate and dried in a laminar airflow. Plates were incubated at 37°C for 24 h and read the next day for results. The same procedures were performed with the standard antibiotic discs used as a positive control, and the average zone of inhibition was recorded (see Table 5.2 and Figure 5.1). The zone of inhibition of plates was compared with a zone of inhibition shown by standard antibiotics by the same bacteria.

Minimum inhibitory concentration (MIC) was determined using the broth microdilution method. In brief, the standard biomass of the test microorganisms, equivalent to 0.5 McFarland and diluted 100 times with meatpeptone broth (MPB), was used for inoculation. The estimated number of microorganisms present was 10^6 colony-forming units (CFU/mL). The CP experimental solution (the concentration range was 0.25–128 mg/L) was prepared from the stock solution (c = 1000 mg/L) using the MPB; the incubation time was 24 h. The MIC was assessed as the lowest concentration, showing no visible bacterial growth (Table 5.3).

5.3 RESULTS AND DISCUSSION

As we can see from Table 5.2, only five polymers (2SpmT, 2SpmC, 2ApdT, 2ApdC, and RSpmCC; polymer sample numbers 5–8 and 10 accordingly) out of the 12 CPs tested revealed the antimicrobial activity against different type of bacterial strains, the zone of inhibition was ranged 14–28 mm. The obtained results showed that the aforementioned five CPs inhibited the growth of the studied microorganisms (Figure 5.1).

The antimicrobial activity of the CPs was quantified as the MIC at which the bacterial growth is completely suppressed after overnight incubation (Table 5.3). The MIC values of CPs against Gram-positive (*B. cereus, S. aureus*) and Gram-negative (*E. coli* and *S. enterica*) bacteria tested are almost in the same range (32–128 mg/L); further, the polymer numbers 5, 7, and 8 revealed the high activity toward all studied microorganisms, while polymer numbers 6 and 10 can penetrate the outer membrane of *Staphylococcus aureus* and *E. Coli* and inhibit their visible growth but these CPs (numbers 6 and 10) did not show any activity against the rest of bacterial strains. The results showed that some of the new CPs suppressed the growth of the microorganisms tested and had selective bactericidal activity. Thus, the selected CP samples can be considered as promising antimicrobial agents, considering their high cytocompatibility.

TABLE 5.1 Summary of Characterization Parameters of the CPs.

Sample number	CP	Class of polymer[a]	Empirical formula (MM)	Yield,% before/ after dialysis[b]	Molecular weight characteristics[c]			Tested for antibacterial activity
					Mn (Da)	Mw (Da)	Dispersity, Đ	
1	1SpmT	PU	$C_{25}H_{40}N_4O_7S_2$ (572.74)	98/38	5,771	22,229	3.85	–
2	1SpmC	PU	$C_{11}H_{26}N_4OCl_2$ (265.18)	100/40	6,632	25,454	3.84	–
3	1ApdT	PU	$C_{13}H_{21}N_3O_4S$ (315.39)	98/45	773	1,168	1.51	–
4	1ApdC	PU	$C_6H_{14}N_3OCl$ (179.65)	100/40	1,109	4403	3.97	–
5	2SpmT	PSuA	$C_{28}H_{44}N_4O_8S_2$ (628.80)	94/47	1,948	3,687	1.89	+
6	2SpmC	PSuA	$C_{14}H_{30}N_4O_2Cl_2$ (321.87)	97/45	595	1,843	3.09	+
7	2ApdT	PSuA	$C_{16}H_{25}N_3O_5S$ (371.45)	100/50	2,417	4,741	1.96	+
8	2ApdC	PSuA	$C_9H_{18}N_3O_2Cl$ (235.71)	95/52	2809	8,340	2.97	+
9	RSpmTC	RPA	$C_{35}H_{59}N_8O_{11}S_2Cl$(867.47)	100/57	2,853	25,945	9.09	–
10	RSpmCC	RPA	$C_{21}H_{41}N_8O_5Cl_3$ (591.96)	99/67	788	2,484	3.15	+
11	RApdTC	RPA	$C_{23}H_{37}N_7O_8SCl$ (607.10)	100/58	1,015	1,519	1.49	–
12	RApdCC	RPA	$C_{16}H_{30}N_7O_5Cl_2$ (435.91)	98/65	956	7,748	8.10	–

[a]*PU, polyureas, PA, polysuccinamdes, RPA, R-attached polyamides (polymalamids).*

[b]*After 4 h dialysis (MWCO 1000 Da).*

[c]*GPC in HFIP, PMMA standards.*

TABLE 5.2 Antimicrobial Activity CPs against Various Bacterial Strains.

##	Test-bacterial strains	Polymer sample* (c = 10 mg/mL)											
		#1	**#2**	**#3**	**#4**	**#5**	**#6**	**#7**	**#8**	**#9**	**#10**	**#11**	**#12**
		Zone of inhibition (mm)**											
1	***B. cereus*** **ATCC 10876**	0	0	0	0	18	0	15	16	0	0	0	0
#2	***S. aureus*** **ATCC 25923**	0	0	0	0	20	28	16	16	0	28	0	0
#3	***E. coli*** **ATCC 8739**	0	0	0	0	18	14	15	14	0	26	0	0
#4	***S. enterica*** **DSM 50912**	0	0	0	0	22	0	13	20	0	0	0	0

**Polymer sample number is corresponding to the numbering given in Table 5.1; for polymer structures see Scheme 5.1.*

***Hole diameter—10 mm.*

TABLE 5.3 MICs of CPs against Various Bacterial Strains.

#	Test-bacterial strain	Polymer sample*				
		#5	#6	#7	#8	#10
		Minimum inhibitory concentration (MIC), mg/L				
1	*B. cereus* ATCC 10876	32	–	>128	8	–
2	*S. aureus* ATCC 25923	64	32	>128	32	32
3	*E. coli* ATCC 8739	64	> 64	128	16	> 64
4	*S. enterica* DSM 50912	64	–	>128	>32	–

**Polymer sample number corresponds to the numbering given in Table 5.1; for polymer structures, see Scheme 5.1.*

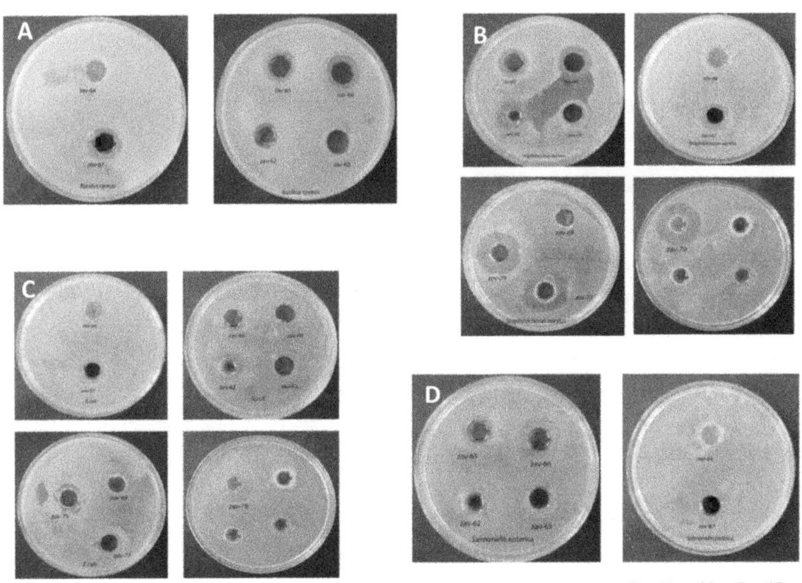

Zav-66 2SpmT (#5)
Zav-75 2SpmC (#6)
Zav-65 2ApdT (#7)
Zav-67 2ApdC (#8)
Zav-79 RSpmCC (#10)

FIGURE 5.1 Antimicrobial activity of CPs* against: A—*Bacillus cereus,* B—*Staphilcoccus aureus,* C—*Esherichia coli*, and D—*Salmonella enterica.*

**Polymer sample number is corresponding to the numbering given in Table 2.1; for polymer structures see Scheme 1.1*

5.4 CONCLUSION

A library of highly water-soluble CPs was designed and synthesized based on the following building blocks: naturally occurring amino acid—arginine (R), endogenous tetraamine—Spm, and synthetic triamine—*N*-(2-aminoethyl)-1,3-propanediamine (Apd). The CPs obtained contain 1,2-diamide fragments in the backbone that provide biodegradation through chemically assisted hydrolysis. The selected CP samples, which had previously shown selective transfection activity against certain cell lines, also showed selective activity against bacteria (tested with *E. coli, S. enterica, S. aureus, B. cereus*). Therefore, considering the high cytocompatibility of the studied CPs, they can be regarded as low-toxic and potentially biodegradable antimicrobials.

KEYWORDS

- **polyamines**
- **spermine**
- **N-(2-aminoethyl)-1,3-propanediamine**
- **arginine**
- **cationic polymers**
- **antimicrobial agents**

REFERENCES

1. Samal, S. K.; Dash, M., Van Vlierberghe, S.; Kaplan, D. L.; Chiellini, E.; van Blitterswijk, C.; Moronid, L.; Dubrue, P. Cationic Polymers and their Therapeutic Potential. *Chem. Soc. Rev.* **2012,** *41,* 7147–7194.
2. Moroson, H.; Rotman, M. Biomedical Applications of Polycations. In *Springer Book Series-Charged and Reactive Polymers: Polyelectrolytes and their Applications*; Rembaum, A., Sélégny, E. Eds.; 1975, Vol. 2, pp 187–195.
3. Kenawy, E.-R.; Worley, S. D.; Broughton, R. The Chemistry and Applications of Antimicrobial Polymers: A State-of-the-Art Review. *Biomacromolecules* **2007,** *8,* 1359–1384.
4. Vaara, M.; Vaara, T. Polycations as Outer Membrane-disorganizing Agents. *Antimicrob. Agents. Chemother.* **1983,** *24,* 114–122.

5. Ikeda, T.; Ledwith, A.; Bamford, C. H.; Hann, R. A. Interaction of a Polymeric Biguanide Biocide with Phospholipid Membranes. *Biochim. Biophys. Acta Biomembr.* **1984,** *769*, 57–66.
6. Morgan, D. M. Polyamins: An Introduction. In: *Polyamine Protocols. Methods in Molecular Biology;* D.M. Morgan, Ed., Vol 79. Humana Press, Inc., Totowa, New Jersey, 1998, pp. 3–30.
7. Grossowicz, N.; Razin, S.; Rozansky, R. Factors Influencing the Antibacterial Action of Spermine and Spermidine on *Staphylococcus aureus*. *J. Gen. Microbiol.* **1955,** *13,* 436–411.
8. David, S. A.; Dutta, A. Polyamines and their use as Antibacterial and Sensitizing Agents. US Patent Appl. Aug 23, 2007. US20070197658 A1.
9. Rabasse, J. M.; De Saegher, J.; Demuynck, M.; Roose, P. Antimicrobial Composition. US Patent appl. Sept 01, 2011. US20110212187 A1.
10. Dewangan, R. P.; Joshi, S.; Kumari, S.; Gautam, H.; Yar, M. S.; Pasha, S. N-Terminally Modified Linear and Branched Spermine Backbone Dipeptidomimetics against Planktonic and Sessile Methicillin-Resistant *Staphylococcus aureus*. *Antimicrob. Agent Chemother.* **2014,** *8* (9), 5435–5447.
11. Ajam, S. M. S.; Salleh, B.; Al-khalil, S.; Sulaiman, S. F. Antimicrobial Activity of Spermine Alkaloids From Samanea Saman against Microbes Associated with Sick Buildings. *Int. Proc. Chem., Biol. Environ. Eng.* **2012,** *49,* 150–155.
12. Samoylenko, V.; Jacob M. R.; Khan S. I.; Zhao, J.; Tekwani, B. L.; Midiwo, J. O.; Walker, L. A.; Muhammad, I. Antimicrobial, Antiparasitic and Cytotoxic Spermine Alkaloids from Albizia Schimperiana. *Nat. Prod. Commun.* **2009,** *4* (6), 791–796.
13. Nezbedová, L.; Drandarov, K.; Deschamps, N.; Werner, C.; Hesse. M. Biogenesis and Function of Macrocyclic Spermine Alkaloids. *ARKIVOC* **2001,** *viii,* 154–164.
14. Kwon, D. H.; Lu, C. D. Polyamine Effects on Antibiotic Susceptibility in Bacteria. *Antimicrob. Agents Chemother.* **2007,** *51* (6), 2070–2077.
15. Jenkins, S.; Addy, M.; Wade, W. The Mechanism of Action of Chlorhexidine. A Study of Plaque Growth on Enamel Inserts in vivo. *J. Clin. Periodontol.* **1988,** *15* (7), 415–424.
16. Graham, J. D.; Lain-Yen, H.; Sharad, M. Therapeutic Substituted Guanidines. *US Patent Appl.* Jul 13, 1999. US5922772 A
17. Manetti, F.; Castagnolo, D.; Raffi, F.; Zizzari, A. T.; Rajamäki, S.; DFigureArezzo, S.; Visca, P.; Cona, A.; Fracasso, M. E.; Doria, D.; Posteraro, B.; Sanguinetti, M.; Fadda, G.; Botta M. Synthesis of New Linear Guanidines and Macrocyclic Amidinourea Derivatives Endowed with High Antifungal Activity Against Candida spp. and Aspergillus spp. *J. Chem. Med.* **2009,** *52,* 7376–7379.
18. Hensler, M. E.; Bernstein, G.; Nizet, V.; Nefzi, A. Pyrrolidine Bis-cyclic Guanidines with Antimicrobial Activity against Drug-resistant Gram-positive Pathogens Identified from a Mixture-based Combinatorial Library. *Bioorgan. Medicin. Chem. Let.* **2006,** *16,* 5073–5079.
19. Infante, M. R.; Pinazo, A.; Molinero, J.; Seguer, J.; Vinardell, P. Arginine Lipopopeptide Surfactants with Antimicrobial Activity. In *Protein-based Surfactants;* Marcel Dekker: New York 2001, *101,* 147–167.
20. Castillo, J. A.; Pinazo, A.; Carilla, J.; Infante, M. R.; Alsina, M. A.; Haro, I.; Clapes. P. Interaction of Antimicrobial Arginine-based Cationic Surfactants with Liposomes and Lipid Monolayers. *Langmuir* **2004,** *30,* 3379–3387.

21. Yeaman, M. R.; Yount N. Y. Mechanisms of Antimicrobial Peptide Action and Resistance. *Pharmacol. Rev.* **2003,** *55,* 27–55.
22. M. Zasloff. Antimicrobial Peptides of Multicellular Organisms. *Nature (London)* **2002,** *415,* 389–395.
23. Cutrona, K. J.; Kaufman, B. A.; Figueroa, D. M.; Elmore, D. E. Role of Arginine and Lysine in the Antimicrobial Mechanism of Histone-derived Antimicrobial Peptides. *FEBS Lett.* **2015,** *589B,* 3915–3920.
24. Xiao, B.; Wan, Y.; Zhao, M.; Liu, Y.; Zhang, S. Preparation and Characterization of Antimicrobial Chitosan-N-arginine with Different Degrees of Substitution. *Carbohydr. Polym.* **2011,** *83,* 144–150.
25. Lahmer, R. A.; Williams, A. P.; Townsend, S.; Baker, S.; Jones, D. L. Antibacterial Action of Chitosan-arginine against *Escherichia coli* O157 in Chicken Juice. *Food Control* **2012,** *26,* 206–211.
26. Tang, H.; Zhang, P.; Kieft, T. L.; Ryan, S. J.; Baker, S. M.; Wiesmann, W. P.; Rogelj. S. Antibacterial Action of a Novel Functionalized Chitosan-arginine against Gram-negative Bacteria. *Acta Biomaterialia* **2010,** *6,* 2562–2571.
27. Mutschler, A.; Tallet, L.; Rabineau, M.; Dollinger, C.; Metz-Boutigue, M.-H.; Schneider, F.; Senger, B.; Vrana, N. E.; Schaaf, P.; Lavalle, P. Unexpected Bactericidal Activity of Poly(arginine)/hyaluronan Nanolayered Coatings. *Chem. Mater.* **2016,** *28,* 8700–8709.
28. Memanishvili, T.; Zavradashvili, N.; Kupatadze, N.; Tugushi, D.; Gverdtsiteli, M.; Torchilin, V. P.; Wandrey, C.; Baldi, L.; Manoli, S. S.; Katsarava, R. Arginine-based Biodegradable Poly(ether-ester) Polymers as Potential Gene Carriers: Cytotoxicity and Complex Formation study *Biomacromolecules* **2014,** *15,* 2839–2848.
29. Kharadze, D.; Memanishvili, T.; Mamulashvili, K.; Omiadze, T.; Kirmelashvili1, L.; Lomtatidze, Z.; Katsarava R. *In Vitro* Antimicrobial Activity Study of Some New Arginine-based Biodegradable Poly(Ester Urethane)s and Poly(Ester Urea)s. *J. Chem. Chem. Eng.* **2015,** *9,* 524–532.
30. Zavradashvili, N.; Sarisozen, C.; Titvinidze, G.; Otinashvili, G.; Kantaria, T.;Tugushi, D.; Puiggali, J.; Torchilin, V. P. and Katsarava, R. Library of Cationic Polymers Composed of Polyamines and Arginine as Gene Transfection Agents. *Acs Omega* **2019,** *4* (1), 2090–2101.

CHAPTER 6

Nuances of the Implementation of the Technological Process in the Mode of Frontal Polymerization in Continuous Conditions

A. O. TONOYAN[1], D. S. DAVTYAN[2], A. Z. VARDERESYAN[1], A. G. KETYAN[1], and N. K. GASPARYAN[1]

[1]*National Polytechnic University of Armenia, Yerevan, RA, Armenia*

[2]*Mckinven Expert LLC, Moscow, RF, Russia*

ABSTRACT

A new approach is proposed to study the nuances of the loss of stability in the process of technological implementation and to detect the loss of linearity and stability of the superabsorbent polyacrylamide hydrogels synthesis in the process of frontal polymerization in continuous conditions (in flow). The implementation of the industrial process in the continuous front mode is carried out in the NPUA "Chemical Technology and Polymer Nanotechnologies" base laboratory, where some deviations from the initial linearity and stability of the front wave under industrial conditions in frontal polymerization (FP) reactors were found with a long process time. In this work, a new approach is proposed and implemented to control stability and establish the nuances of microkinetic disturbances due to temporary disturbances in the ambient temperature, as well as in the velocity and temperature of the supplied flow.

Advances in Engineering Materials: Preparation for Sustainable Process Development. Tamara Tatrishvili and A. K. Haghi (Eds.)

6.1 INTRODUCTION

Frontal polymerization (FP) is engaged in many prominent research institutes and universities around the world.[1–8] The FP method is nontraditional, ecologically safe, and energetically efficient in the mode of FP wave. Owing to the unique characteristics of this method and its vast application potential in high-level technologies, research involving various monomers and conditions has proliferated across numerous countries. Numerous studies have also been carried out on the synthesis of polyacrylamide hydrogel (PAH) in the FP mode under periodic conditions,[9–11] which is of great scientific interest, but not provided for practical support. We continue to conduct batch studies to generate foundational data tailored to the specific requirements for the synthesis of a particular polymer and transfer the resulting data to a continuous process in flow FP mode. Until now, the production of FP under continuous conditions has been carried out only by us; now, research is carried out on the industrial process for the synthesis of PAH in the FP mode in flow.[12–14]

Currently, our research related to the nuances of regulating kinetic and macrokinetic parameters for the synthesis of PAH by the FP method is being conducted, both theoretically and experimentally, using the scientific foundations of calculations and models carried out during the formation and development of the FP method.[1,4,5,15–19] In this process, it is vital to consider the nuances of the process in a continuous FP mode during technological implementation, in addition to the main parameters of the kinetics and macrokinetics of the process, as well as the nuances of the system exiting the stationary state as a result of a violation of the stability of the kinetics of the elementary reaction.

The processes of nonisothermal polymerization in the frontal mode are similar in their kinetic laws to adiabatic processes, with the only difference of the front adiabatic temperature curves repeating according to the propagation of the front wave.[1] Having the laws of polymerization in the adiabatic regime, we can calculate the kinetics and mechanism in the front, which was fundamental in S.P. Davtyan's and A.O. Tonoyan's created works for FP kinetics models and calculations.[1,4,5,15–19] We use it to establish the nuances of the wave propagation of the FP synthesis of PAH in the phase transition mode of continuous supply of monomer during industrial implementation, taking into account the nuances of the loss of stability and linearity of the propagation of the front wave in continuous reactors. FP method for the synthesis of PAH is of particular interest. PAH are three-dimensional cross-linked polymers. These gels are in demand in various areas of the national economy due to their ability to absorb huge amounts of water and return the absorbed water without changing their structure and properties. Their usage field is vast: starting from agriculture, pharmacology, medicine, the oil industry, removing poisonous liquids from garbage cans, and ending with personal hygiene items—diapers, etc.

PAH synthesis by FP was not previously implemented by anyone, but we could achieve it. The process becomes macrokinetically controlled and profitable when continuous conditions are created in industrial-type FP reactors.[16] Previously, we have implemented industrial FP reactors in continuous conditions for liquid monomers and introduced them into the industry.[20] The experience allowed us to develop an FP reactor for PAH synthesis.[12]

6.2 RESULTS AND DISCUSSION

FP is a process in which the heat required to initiate the polymerization process is supplied to the first layer of the initial mixture, and the heat source is immediately removed. This heat causes an exothermic polymerization process in this layer; the released heat is transferred to the adjacent layer of the initial mixture, causing the next layer of polymerization. Thus, the FP process from one edge of the reactor goes to the end in the autowave mode of thermal wave propagation.

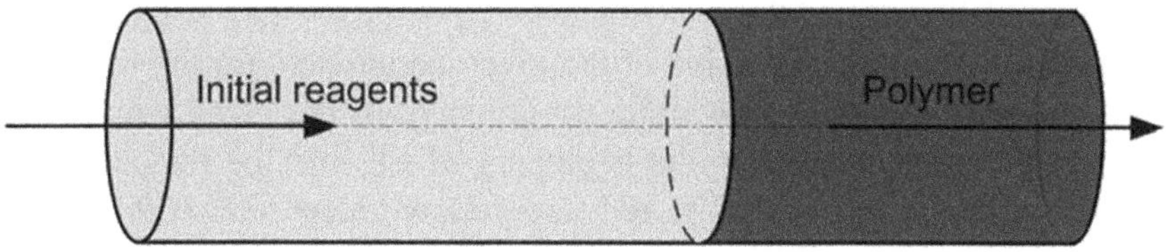

FIGURE 6.1 PAH synthesis by FP in a continuous reactor.

For the FP process in tubular geometry, the establishment of a stable stationary front occurs in the direction of the front propagation not only along the length of the reactor but also along the radius when transitions of autowave propagation from one state to another are possible depending on various kinetic parameters of the reaction and macrokinetic influence on the FP process. We studied both the impact of flow rate, kinetics and adiabaticity of the environment, and macrokinetic parameters in continuous reactors, and a little deviation in reactions along the radius or the length of the reactor during the continuous process of supplying monomer and removing the resulting polymer, deviations from frontal stability are possible, which can affect the properties of the resulting PAH with impaired synthesis reproducibility. In the process of continuous FP, violations of the linearity and stability of the technological process over time were found. The calculations were carried out using numerical methods in comparison with experimental studies.[16,18,19]

Violations of the linearity and stability of the technological process over time of this reaction were detected by numerical and experimental studies. Calculations in this direction were carried out in the 3D dimension in the process of a continuous front, where a clear visual opportunity is created to follow the nuances of the process. Numerical calculations in 3D measurement have played an important role in controlling the stationarity of FP. Figure 6.2 presents the 3D images for different macrokinetic conditions of PAG synthesis with the regulation of not only macrokinetic parameters but also taking into account the elementary acts of nonisothermal processes researched under adiabatic conditions with application to the front. For the FP process in a tubular reactor, the stationarity of the front both along the length and radius of the reactor is influenced by various kinetic parameters and macrokinetic factors, and any change in the speed of wave propagation affects the properties of the resulting product.

In some places in Figure 6.2, along the length of the tubular reactor and in the radial direction of the front, there are violations of the FP stationarity. Accordingly, the average value of the given parameters for the resulting hydrogel may fluctuate with some deviation from the given value. The region of monomer heating, the beginning of FP, and the region of the formed polymer are marked in red, the yellow-green zone is the stable front, and the zone of the initial mixture is blue.

Figure 6.3 shows that during hydrogel synthesis, depending on the propagation range process in a continuous frontal mode, after a certain period, a jet of flow can form, violating the uniformity of autowave propagation in the radial direction. Depending on the temperature on the reactor walls, the curves can be linear, where the amplitude and phase of the wave along the radius and the length of the reactor flow stably with linear wave propagation and the same adiabatic wave temperature. However, a violation of the heat supplied from outside can also increase FP speed at the walls and a decrease in the wave height in the direction of the flow supply; at the same time, the high temperature of the reactor walls can lead to burnout of the initiator, which will lead to the weakening of the FP at the walls, decreasing length and height of the wave, while in the center of the flow, there will be increasing of height and speed FP wave propagation.

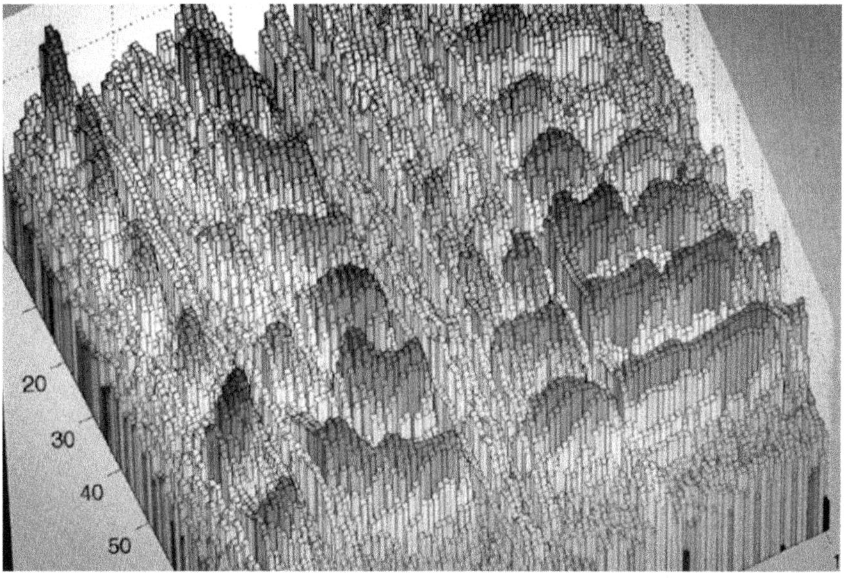

FIGURE 6.2 Distribution of the FP in continuous conditions due to the flow of the initial mixture and the output of the final product in a continuous mode.

The comparison of the experiment with calculations gives the opportunity to exclude small deviations from the stationary and nonlinearity of the production process, which may affect the reproducibility of the obtained materials. Here are figures that allow the establishment of violations of linearity when the ambient temperature changes with a violation of the adiabatic process and changes in the temperature and flow rate of the initial mixture.

Below are the images for calculating the FP of PAG for different macrokinetic conditions carried out in a pilot plant in the flow of initial substances. In the numerical solution of the heat equation, an implicit four-point scheme was used since the considered values of temperature and concentrations had a sharp change in a relatively narrow reaction zone (Figure 6.4). Particular attention was paid to the value of the variable step in space. When solving the problem, dimensionless time, coordinate, and concentration were introduced. The initial and boundary conditions were taken as *T* (zero) where *X* is from 0 to 1 and *T* (ignition) where *X* is more than 0 and up to 1+, additional parameters, changing adiabatic process introduced (for example, a change in ambient temperature), the ambient temperature, the temperature of the initial mixture, the ignition temperature and the feed rate of the initial mix were varied. In the figures, the heating area of the sample layer of the polymerizing mass is marked in red. The next yellow area corresponds to the FP area, and the below part of the initial mixture is colored blue. For mathematical calculations, a narrow wavefront propagation zone is used, where the temperature profiles are close to isothermal ones, and calculations for quasi-stationary concentrations can be carried out by analogy with isothermal polymerization. A 3D image of the front in a narrow zone is presented in Figure 6.4.

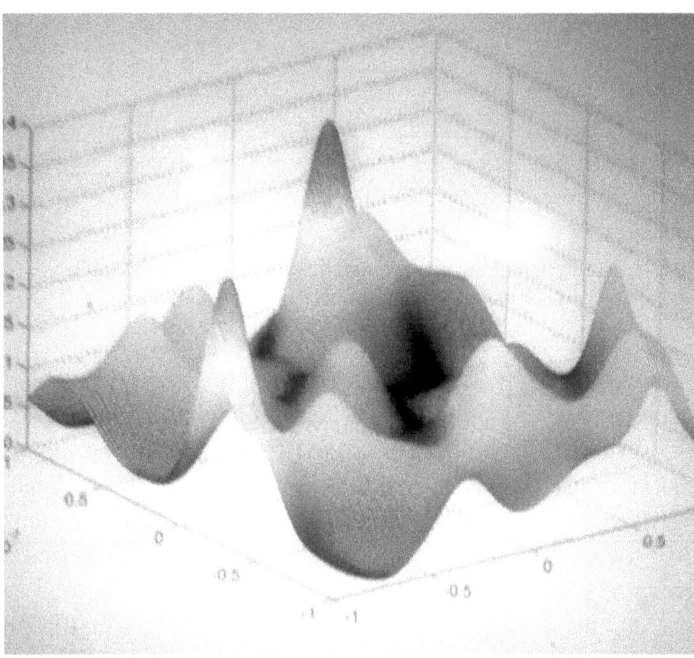

FIGURE 6.3 The increase of ambient temperature by 20° violates the adiabaticity, which leads to the loss of stationarity of FP.

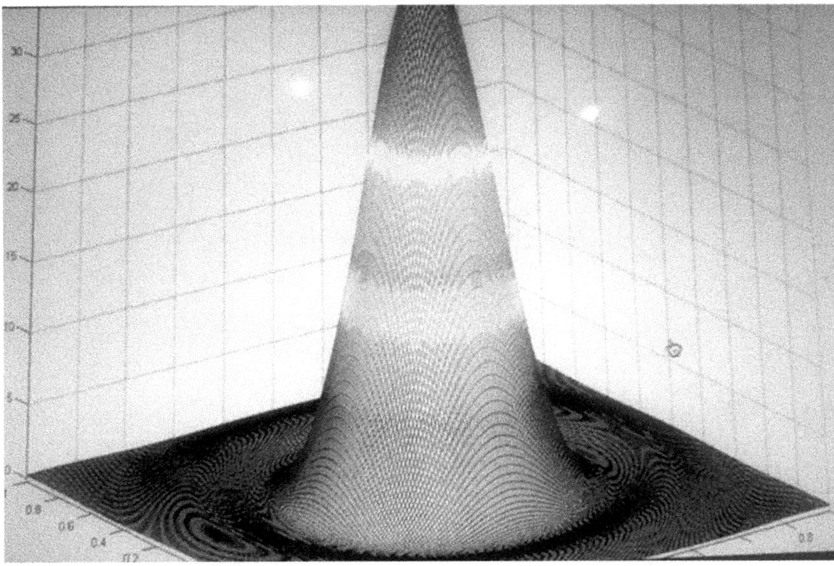

FIGURE 6.4 3D image of the front in a narrow zone.

In the process of industrial implementation, different bursts of the FP were found both along the reactor's length and the radius in time. Adhesion of the polymer on the reactor's walls with the formation of a jet of monomer through the polymer melt can occur both as a result of the dissipation heat release from the friction of the polymer flow with the reactor's walls and the reflection of changes in ambient temperature on the adiabaticity of the process.

6.3 CONCLUSION

Studies of various features of PAH synthesis make it possible to create a real basis for the practical synthesis of PAH technology. We studied both the influence of flow rate, kinetics and adiabaticity of the environment, and macrokinetic parameters in continuous reactors, and a little deviation in reactions along the radius or the length of the reactor during the continuous process of supplying monomer and removing the resulting polymer, deviations from frontal stability are possible, which can affect the properties of the resulting PAH with impaired synthesis reproducibility.

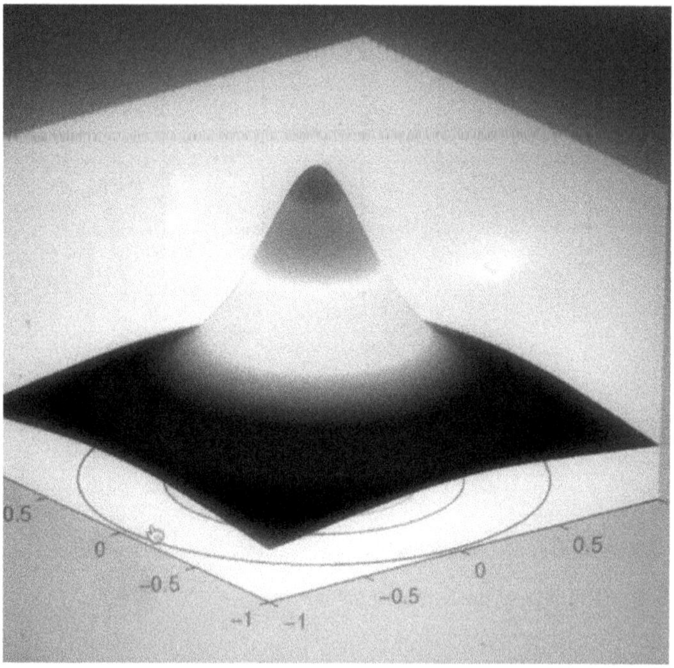

FIGURE 6.5 3D image of the front in a wide zone.

Compared to Figure 6.4, the front zone is wide.

KEYWORDS

- **frontal polymerization**
- **polyacrylamide hydrogel**
- **linearity loss**
- **continuous reactors**
- **3D images**

REFERENCES

1. Davtyan S. P.; Tonoyan A. O. Theory *and Practice of Adiabatic and Frontal Polymerization*; Palmarium Academic Publishing: Germany, 2014; p. 660.

2. Davtyan, S. P.; Arutyunyan, K. A.; Shkadinskii, K. G.; Rozenberg, B. A.; Yeniko-lopyan, N. S. The Mechanism of Epoxide Oligomer Hardening by Diamines Under Advancing Reaction Front Conditions. *Polym. Sci. USSR* **1978,** *19,* 3149–3154.
3. Babadzhanyan, A. S.; Volpert, V. A.; Volpert. V. L. A.; Davtyan, S. P.; Megrabova, I. N. Frontal Regimes of an Exothermic Reaction with Radially Symmetric Injection of Reagents. *Comb. Expl. Shock Waves* **1988,** *24,* 711–7194.
4. Davtyan, S. P.; *Monography, Non-isothermal Methods of Polymer Synthesis. Theory and Practice of Adiabatic Polymerization*; Yerevan. 2004; p 320.
5. Davtyan, S. P.; Tonoyan, A. O. The Frontal Polymerization Method in High Technology Applications ISSN 2079-9780. *Rev. J. Chem.* **2019,** *9* (1), 71–94.
6. Solovyov, S. E.; Volpert, V. A.; Davtyan, S. P. Radially Symmetric Flow of Reacting Liquid with Changing Viscosity. *SIAM, J. Appl. Math.* **1993,** *53,* 907–914.
7. Volpert, V.; Volpert, V. A.; Pojman, J. A.; Solovyov, S. E. Hydrodynamic Stability of a Polymerization Front. *Eur. J. Appl. Math.* **1996,** *7,* 303–320.
8. Goldfeder, P. M.; Volpert. V. A. A Model of Frontal Polymerization Using Complex Initiation. *Math. Prob. Eng.* **1999,** *5,* 139–160.
9. Washington, R. P.; Steinbock, O. Frontal Polymerization Synthesis of Temperature-Sensitive Hydrogels. *J. Am. Chem. Soc.* **2001,** *123,* 7933–7934.
10. Lu, G. D.; Yan, Q. Z.; Ge, C. C. Preparation of Porous Polyacrylamide Hydrogels by Frontal Polymerization. *Polym. Intern.* **2007,** *56,* 1016–1020.
11. Sanna, R.; Sanna, D.; Alzari, V.; Nuvoli, D.; Scognamillo, S.; Piccinini, M.; Lazzari, M.; Gioffredi, E.; Malucelli, G.' Mariani, A. Synthesis and Characterization of Graphene-Containing Thermoresponsive Nanocomposite Hydrogels of Poly(N-vinylcaprolactam) Prepared by Frontal Polymerization. *Sci., Part A: Polym. Chem.,* **2012,** *50,* 4110–4118.
12. Davtyan, D. S.; Tonoyan, A. O.; Davtyan, S. P. RU Patent N2018124992, 2019.
13. Davtyan, S. P.; Tonoyan, A. O.; Gevorgyan, L. A.; Varderesyan, A. O.; RA Patent N 2697A, 2012.
14. Davtyan, S. P.; Tonoyan, A. O.; Gevorgyan, L. A. RA Patent no. 2235A, 2008.
15. Tonoyan, A. O.; Davtyan. S. P.; Muller, S. C.; *Bottom-Up Self-Organization in Supramolecular Soft Matter. Principles and Prototypical Examples of Recent Advances*; Influence of nanoparticles on the mechanism and properties of nanocomposites obtained in frontal regime. Springer Series in Materials Science, 2015, Vol. 217. Chapter 5, pp 101–126.
16. Tonoyan, A. O.; Varderesyan, A. Z.; Ketyan, A. G.; Minasyan, A. H.; Hovnanyan, K. O.; Davtyan, S. P. Continuous Reactors of Frontal Polymerization in Flow for the Synthesis of Polymacrylamide Hydrogels with Prescribed Properties. *J. Polym. Eng.* **2020,** 40 (7), 601–606., DOI: https://doi.org/10.1515/polyeng-2019-0369.
17. Tonoyan, A. O.; Minasyan, A. H.; Varderesyan, A. Z.; Ketyan, A. G.; Davtyan, S. P. *Investigation of the Causes of Stability Violation of Propagating Polymerization Heat Waves in the Process of Frontal polymerization. Advanced Materials, Polymers and Composites: New Resarch on Properties, Techniques and Applications. Part 1: Composites and Nanomaterials*; Apple Academic Press, 2021; Chapter 20, pp 289–296.
18. Davtyan, S. P.; Tonoyan, A. O. Frontal Polymerization in Continuous-Flow Reactors. Rev. J. Chem. **2019,** *9* (4), 175–196.

19. Tonoyan, A. O.; Minasyan, A. H.; Varderesyan, A. Z.; Ketyan, A. G.; Davtyan. S. P. Influence of Shrinkage of Polymer on the Stationarity of Frontal Polymerization Heat Waves. *J. Polym. Eng.* **2019,** *39* (8), 769–773.
20. Davtyan, S. P.; Tonoyan, A. O.; Radugina, A. A.; Davtyan, D. S.; Abrosimov, A. F.; Savchenko, V. I. Frontal Radical Polymerization of Methyl Metacrilate in a Cylindrical Flow Reactor. Vysokomol. Soed. **1999,** *41* (2), 232–241.

CHAPTER 7

Study of Pellets Obtained from Biomass

KAKHA KARCHKHADZE, NATELA KHETSURIANI, ELZA TOPURIA, IRINA MCHEDLISHVILI, and MADLENA CHKHAIDZE

Petre Melikishvili Institute of Physical and Organic Chemistry of the IV. Javakhishvili Tbilisi State University, Tbilisi, Georgia

ABSTRACT

One of the most important problems of the world's energy sector is to develop new, innovative technologies that can save natural resources, which are not unlimited. One of the most important tasks in this field is the research of alternative, renewable sources of energy. These energy sources have additional advantages if/when they are based on the processing of renewable biomass, secondary raw materials, and agricultural wastes. The production of fuel pellets from waste biomass meets all these criteria.

The main goal for us was to prepare and study the properties of combustible pellets based on different compositions using the biomass wastes of energy-bearing raw materials available in Georgia; among them were the by-products remaining after processing the fir tree cones and the maize (corn) roots as bio-residues. We carried out testing the samples of the pellets made from these raw materials and analyzed their physical-chemical, mechanical, and technical characteristics. Combustible pellets from Caucasian fir cones, as well as samples of the mixture of 15 and 35% with sawdust, were prepared and studied to use them as alternative heating materials.

For the first time in Georgia's energy sector, the Caucasian fir cones were used for the production of pellets. It is estimated that Caucasian fir

Advances in Engineering Materials: Preparation for Sustainable Process Development. Tamara Tatrishvili and A. K. Haghi (Eds.)

cones contain about 37% polysaccharides, 40% lignin and 7% polyphenolic compounds, 13% of lipids, 12% of protein compounds, and less than 1% of essential oils. The cones and the seeds of this species also contain micro and macro elements, which determine their physical properties important for the pellet's production.

We have analyzed the properties of the obtained pellet samples and found that the calorific values were within 15–17 MJ/kg, with a humidity of 10–12%. The ash content varied within 3–4%, and stability varied between 60 and 95 kg/cm^2. The high value of stability was caused by the natural glue contained in the composition. The pellets burnt four times longer than the original material and the pellets also produced a relatively small amount of smoke during burning, thus they had no negative impact on the environment. Overall, we found that the physical and chemical properties of the samples of the pellets complied with the requirements of the international standard.

The results of our research demonstrated that through the processing of renewable sources of biomass and various agricultural wastes available in Georgia, the industry can develop and manufacture high-calorific alternatives to firewood, pellets/briquettes, that is, solid fuel.

7.1 INTRODUCTION

The last decade was marked by a significant increase in primary energy consumption worldwide, which reached 13.5 billion tons of oil equivalent totaling about 565 EJ (exajoules), and at the same time, the decadal average growth rate was estimated as 1.7% per year.[1]

The consequence of ever-increasing energy needs, depleting fossil fuel reserves, and raising environmental concerns are making the consumption of renewable energy inevitable and important for both the economy and the environment.

The Paris Agreement of 2015 gave a new impulse to the development of the renewable energy market, implying a reduction of greenhouse gas emissions related to the usage of fossil hydrocarbons. Since the Paris Agreement is a binding treaty for all undersigned nations, it will reduce the usage of fossil hydrocarbons and fight global warming, in order to maintain the rise in global temperature under 2°C compared to the pre-industrial level.[2]

The United Nations also emphasized the role of renewable energy by including it in the list of sustainable development goals (SDGs), namely,

"Affordable and clean energy" (SDG #7), as well as "Climate action" (SGD 13). All of these highlight the significance of alternative, clean energy sources, and renewable energy.[3] In this context, energy sustainability goes hand-in-hand with sustainable development. Renewable, clean, and affordable energy supply is the core of sustainable energy for the modern world and future development.[4] Renewable, clean energy will continue to play an important role in the world's future energy mix.

As per REN21's 2017 report by the end of the second decade of the 21st century, renewable energy was contributing about 19.3% to the final energy consumption worldwide.[5] Solar, wind, hydro, biomass, and geothermal are major sources to generate renewable energy. Of the total share of the renewable energies, the share of biomass energy is the largest and accounts for 9% (~51 EJ) of the total primary energy supply in the world—and out of that approximately 55% is related to traditional usages, such as cooking and heating.[6]

Slade et al.[7] analyzed the potential of biomass and concluded that biomass can contribute up to ~100 EJ in the world energy supply, of which 70 EJ is from energy crops and ~30 EJ from wastes. Since biomass is considered a carbon-neutral fuel, the usage of biomass does not add carbon dioxide to the atmosphere, which is a clear advantage of renewable biomass over fossil fuels.[8]

The agricultural residues and other organic wastes after processing can substitute fossil fuels and reduce emissions. In this context, biomass obtained from the agricultural sector has significant potential to be used as a sustainable source of renewable energy and contribute to the further production of energy from renewable sources.

7.2 EXPERIMENTAL METHODS AND MATERIALS

A typical fuel pellets and briquettes production process consists of a series of pre-processing steps depending on the initial biomass characteristics. Pre-processing plays an important role in a pelletization system as it improves bulk density, energy density, storage, and handling characteristics of biomass. Some common pre-processing methods such as size reduction, torrefaction, steam explosion, hydrothermal carbonization, and biological treatment are discussed in this section. The basic scheme of pellets/briquettes production is given in Figure 7.1.

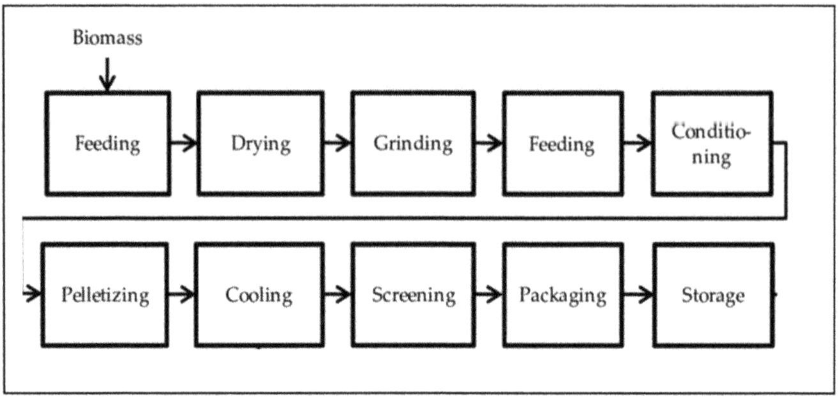

FIGURE 7.1 Schematic of the production process of biomass pellets/briquettes.

In our research, we focused on two important sources of biomass available in Georgia, namely, Maize (corn) roots and the lower part of the stem that remained after corn harvest, and coniferous three cones, namely, fir tree cones' residue materials that remain after processing of the cones.

These two sources have great potential to provide thousands of tons of raw materials for producing pellets and briquettes in Georgia. The Georgian fir tree (Latin: *Abies nordmanniana*) is an excellent tree for the Christmas tree business in Europe. Every year about 100 tons of seed are exported to the EU countries, which means every year about 1000 tons of cones are processed and bio waste is left as a result of processing the fir tree cones.

Our research was focused on the study of these waste materials and our goal was to find out, but not limited to, whether they would be suitable for producing pellets and briquettes. We used several other components such as clay and sawdust, as well as some chemicals while preparing the samples of solid fuel, that is, pellets and briquettes.

The raw materials of these two main feedstocks were dried, chopped, and prepared for pellet and briquette production. Along with that, we prepared several (various) composites to increase caloric content and the quality of the burning process.

As the bio or organic wastes, we used the waste left after processing timber materials of pine and fir trees and the sawdust of these trees.

The timber reserves in Georgia are estimated at 951.7 million cubic meters. Out of that the amount for processing is only 40%. For the production of pellets and briquettes, the following waste materials may be used: sawdust, mulch, chops, bark, cones, and cone wings. The average heat capacity for various species of these species is 4753 kcal/kg The properties of these materials are essential for the industry for solid fuel production therefore we analyzed their parameters, which are given in Table 7.1. which shows the properties of the timbers of the main tree species widely spread in Georgia.

TABLE 7.1 The Forestry Fund of Georgia.

Sample	Area 1000 hectare	Timber reserves 1000 hectares	Volumetric heat capacity kcal/kg	Wood density kg/dm^3
Oak	186	17,525.42	3240	0.690–1.03
Beech	1035	2,365,668.84	3000	0.620–0.820
Pine	2005	11,201.40	2080	0.310–0.760
Spruce	1195	34,660.85	1800	0.370–0.750
Chestnut	46.6	10,529.88	2600	0.600–0.720
Fir	850.8	2,154,775	4880	0.345–0.655

As we said earlier, the clays of various deposits were used for giving a suitable form to the final product—solid fuel, and as a modification, we used manganese oxide, magnesite, and calcium carbonate.

While preparing the samples of the solid fuel, that is, briquettes and pellets, we used various clays to give the final form to the bio-waste mixture. To activate the burning process, we used Urotropine and some other components. We have also used geological substances containing manganese, manganese di-oxide, and magnesite.

For production purposes, the availability of the composite materials is very important. Therefore, to fill the burning composites, we used clays that are widely spread in Georgia[9] such as Askana clays with a reserve estimated at 9.5 mln tons, and Gumbrine—5.2 mln tons.

Figure 7.2 shows the chemical composition of the clay of Askana, Gumbrin, and Gareji deposits.

From the samples of the solid fuel prepared, we have selected three bio and four complex compositions that have properties that meet the standards. The amount of harmful gas emissions was identified by air–liquid

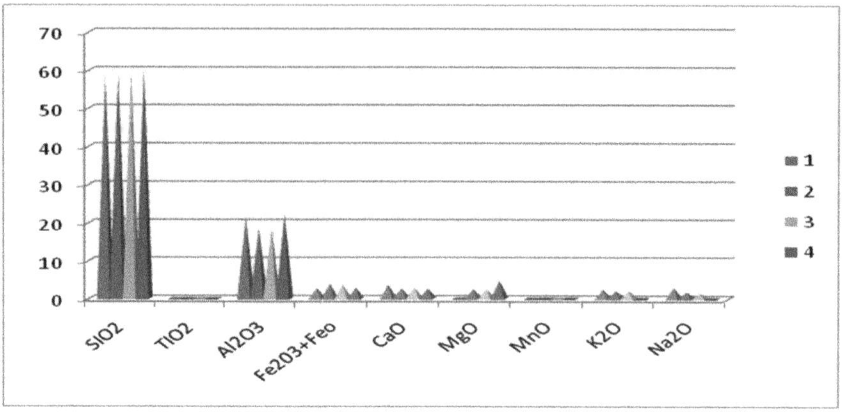

FIGURE 7.2 Chemical analyses of the clays of Askana, Gumbrine, and Gareji deposits.

chromatography analyses. The results and the energy released are given in Table 7.2.

We measured the physical-chemical and technical-mechanical properties of the burning materials and mixed pellets/briquettes and then compared them—the results are given in Table 7.3.

TABLE 7.2 The Gas Emissions During the Burning of the Solid Bio-Fuels.

#	Sample name	Heat capacity kcal/kg	Exhaust gases, M^3/kg			The energy released by combustion	
			SO_x	NO_x	CO_x	MJ/kg	kWh
1.	Industrial saw dust	3956	0.8454	0.0017	0.0038	16.56	4.60
2.	Corn root powder	3868	0.8210	0.0065	0.0002	16.19	4.50
3.	Sawdust	3955	0.8379	0.0015	0.00014	16.56	4.60
4.	Composition 1	4878	0.9409	0.0084	0.0039	20,42	5.67
5.	Composition 2	5570	0.9982	0.0068	0.00413	23,32	6.48
6.	Composition 3	5355	0.9769	0.0169	0.0084	22.42	6.23
7.	Composition 4	5321	0.9698	0.01703	0.0077	22.28	6.19

We have found out that the physical-chemical properties of the pellets and briquettes made of maize roots, sawdust, and fir three cones burn without smoke and thus have minimal harmful impact on the environment. In addition, the remaining material, after their burning, contains Kalium, which can be used as fertilizer. In Table 7.3, there are given standards for wooden pellets.

We have found out that the composites have higher calory (thermal) content compared to the raw materials. The calory content of the solid fuel reaches 4900–5600 kcal/kg; therefore, they can be used in the energy industry, in thermal power plants, greenhouses, buildings, and various enterprises.

An extra advantage of biofuel compared to firewood is the fact that the emissions while burning solid biofuels are within the related standards. The composites and solid fuel give clear ecological benefits. In order to evaluate the ecological benefits, we have calculated emissions (m^3/kg, m^3/kcal, m^3/kWh), as well as the coefficients g/mJ of the emissions of CO_2, SO_2, and NO_2. The results are given in Tables 7.3. and 7.4.

Characteristics of the obtained pellets were analyzed. We received the following results:

- Calorific value 16.5–17.2 MJ/kg,
- Moisture 9–10%,
- Ash content varies within 2.5–3.0%,
- Durability (min 95–102 kg/cm^2).

The main parameters of the pellets met the EU standards and demands.

7.3 RESULTS AND DISCUSSIONS

- The physical and chemical characteristics of the obtained pellets were analyzed. The results obtained are: calorific value 16–17.2 MJ/kg, humidity 10–12%, ash content varies within 3–4.4%, and stability (60–95 kg/cm^2). Stability seems to be given by the glue contained in the composition. The burning duration of the pellets where 4.5 times higher than that of the original material.
- These pellets produced a smaller amount of smoke released during burning to compare the raw materials; thus, they have a less negative impact on the environment. Their mechanical and technical characteristics comply with the requirements of the standard. The further advantage is that the ash that remained as a result of pellets burning is the best potassium-containing fertilizer.
- Combustible pellets from Caucasian fir cones, as well as from their compositions with 15 and 35% sawdust, were developed and

TABLE 7.3 The Properties of Raw Materials, Organic Waste Fuels, and Mixed Composites.

Raw materials, pellets, composites	Density g/sm^3	Moisture %	Ash content %	Emissions m^3/kg			Energy received		
				CO_x	NO_x	SO_x	Heat content kcal/kg	mJ/kg	kWh
Maize roots grinded	1.41	3.00	5.1	0.8210	0.0065	0.0002	3868	16.19	4.50
Saw dust (Pine +Spruce)	1.28	8.10	1.17	0.8379	0.0015	0.0001	3955	16.56	4.60
Composite 1	1.30	5.88	12.46	0.9409	0.0084	0.0039	4878	20,42	5.67
Composite 2	1.23	5.05	10.06	0.9982	0.0068	0.0041	5570	23,32	6.48
Composite 3	1.40	2.47	26.87	0.9769	0.0169	0.0084	5355	22.42	6.23
Composite 4	1.34	2.46	27.22	0.9698	0.01703	0.0077	5321	22.28	6.19

**The samples were made under the following conditions*: temperature 100–250°C; and the pressure 250 kg/cm^2.

TABLE 7.4 Evaluation of Ecological Benefits from Solid Fuels.

#	Raw materials, pellets, composites	Heat content kcal/kg	Emissions m^3/kg			Emissions M3/kWh.10-3		
			CO_x	NO_x	SO_x	CO_x	NO_x	So_x
1.	Timber dust	3956	0.8454	0.0017	0.0038	183.780	0.3695	0.8260
2.	Ground maize	3868	0.8210	0.0065	0.0002	182.440	1.4444	0.0444
3.	Saw dust	3955	0.8379	0.0015	0.0001	182.215	0.3260	0.0304
4.	Composition 1	4878	0.9409	0.0084	0.0039	165.94	1.4814	0.6878
5.	Composition 2	5570	0.9982	0.0068	0.0041	154.04	1.0493	0.6373
6.	Composition 3	5355	0.9769	0.0169	0.0084	156.80	2.7126	1.3483
7.	Composition 4	5321	0.9698	0.0170	0.0077	156.67	2.7512	1.2439

TABLE 7.5 Quality Parameters of Wooden Pellets.

Quality parameter	**Unit**	**ENPlus-A1**	**ENPlus-A1**	**EN-B**
Diameter	mm	6 or 8 ±1	6 or 8 ±1	6 or 8 ±1
Length	mm	3.15–40	3.15 40	3.15 40
Bulk density	kg/m^3	>600	> 600	>600
Net calorific value	MJ/kg	16.5–19	16.5–19	16.5–19
Moisture	%	<10	<10	<10
Durability	%	>97.5	>97.5	>97.5
Ash	%	<0 <7	<1,5	<3
Ash melting temp.	°C	>1200	> 1100	> 1100

manufactured for heating various purpose buildings such as small boilers, production, energy, household stoves, and fireplaces.

7.4 CONCLUSIONS

As the results have demonstrated, it is possible to develop and obtain high-calorific alternative, solid biofuel based on the utilization of renewable biomass sources and agricultural wastes available in Georgia, which can contribute toward the development of energy-efficient technology and reduce the environmental burden.

The produced samples of pellets fully meet the standard requirements and can benefit the economy of the country and contribute towards the environmental protection of Georgia.

The solid biofuel successfully can be used to provide additional thermal energy, which opens new possibilities for the development of the relevant industry sector in the country.

KEYWORDS

- **biomass**
- **pellets**
- **solid biofuel**
- **emissions**

REFERENCES

1. British Petroleum UK. BP Statistical review of world energy [Online], 2018. https://www.bp.com/en/global/corporate/energy-economics/statistical-review-of-world-energy.html (accessed Sep 22, 2018).
2. UNFCCC. United Nations Framework Convention on Climate Change [Online], 2018. http://unfccc.int/paris_agreement/items/9485.php (accessed Mar 26, 2018).
3. UNSDG. United Nations Sustainable Development Goal report [Online], 2018. http://www.un.org/sustainabledevelopment/ (accessed Mar 26, 2018).
4. UNGA. World Commission on Environment and Development: Our Common Future. Oslo, Norway: United Nations General Assembly, Development and International Cooperation: Environment [Online] 1987, http://www.un-documents.net/our-common-future.pdf (accessed Apr 21, 2018).
5. REN21 Renewables global status report, REN21 Secretariat, Paris [Online], 2017. http://www.ren21.net/wp-content/uploads/2017/06/17-8399_GSR_2017_Full_Report_0621_Opt.pdf (accessed Jun 21, 2018).
6. IEA International Energy Agency, Technology Roadmap: Delivering Sustainable Bioenergy. OECD/IEA Paris [Online] 2017. http://www.iea.org/publications/freepublications/publication/Technology_Roadmap_Delivering_Sustainable_Bioenergy.pdf (accessed Jun 21, 2018).
7. R. Slade, A.; Bauen, R. Gross, Global Bioenergy Resources *Nat. Clim. Chang.* **2014,** *4,* 99–105. DOI:10.1038/nclimate2097.
8. Saidur, E.A.; Abdelaziz, A.; Demirbas, M.S.; Hossain, S. Mekhilef A Review on Biomass as a Fuel for Boilers. *Renew. Sust. Energ. Rev.* **2011,** *15,* 2262–2289. DOI: 10.1016/j.rser.2011.02.015
9. https://www.bp.com/content/dam/bp/business-sites/en/global/corporate/pdfs/energy-economics/energy-outlook/bp-energy-outlook-2019.pdf

CHAPTER 8

Investigation of the Regularities and Properties of the Oxidative Copolycondensation Reaction Products of 1-Naphthylamine with 1-Naphthol

GUNEL ABASZADE, RENA AKHMEDOVA, CHIMNAZ ISMAILOVA, SHABNAM GUSEINOVA, BAKHTIYAR MAMEDOV, and DILBAR ALIYEVA

Institute of Polymer Materials of the Ministry of Science and Education of the Republic of Azerbaijan, Sumgait, Azerbaijan

ABSTRACT

The oxidative copolycondensation reaction of 1-naphthylamine with 1-naphthol in the presence of an aqueous solution of hydrogen peroxide has been carried out. The results of the investigation showed that the temperature rises in the range of 343–368 K, the molar ratio of H_2O_2:(NA+Nph) in the range of 1:1–3:1, and also the quantity of naphthylamine in the reaction mixture leads to an increase of the yield of macromolecular compounds as a result of the interaction of 1-NA with 1-Nph in the presence of H_2O_2. For this process, the optimal value of the reaction duration is 8 h and the addition of ethanol to the reaction medium leads to an increase in the yield of co-oligomers. On the basis of the gel-chromatogram of the samples of cooligomers synthesized under various conditions, their molecular weight indices have been found. It should be noted that the curves of the molecular weight distribution of the oxidative polycondensation products of NA and Nph have a bimodal character. The reaction products consist of low-molecular $\bar{M}_n = 330 - 580$ and high-molecular $\bar{M}_w = 5230 - 6480$ fractions.

Advances in Engineering Materials: Preparation for Sustainable Process Development. Tamara Tatrishvili and A. K. Haghi (Eds.)

The growth of the quantity of H_2O_2 oxidant and the reaction temperature is accompanied by an increase of molecular weight indices and simultaneously polydispersity of the cooligomers. The synthesized oligomers of 1-naphthylamine with 1-naphthol (CONN) have been used as an active additive for the preparation of rubber mixtures on the basis of butyl rubber (BR). CONN samples exhibit high electrical conductivity; their joint use with electrically conductive carbon black allows to obtain rubbers with σ_v $\sim 10^{-8}$–10^{-6} $Om^{-1} \cdot cm^{-1}$. The growth of CONN content from 22.5 to 45.0 m.p. (from the mass of rubber) instead of carbon black leads to a decrease in the specific resistance of the obtained rubbers. The percolation effect is observed at −23.6 m.p. of CONN for rubbers obtained from BR. CONN can also act as a function of an antioxidant in compositions.

8.1 INTRODUCTION

The preparation of electrically conductive or antistatic rubbers, widely used in various fields of engineering and industry, is possible with the use of electrically conductive fillers.[1] The metallic powders, various marks of carbon black, and powdered graphite are usually used as electrically conductive components.[2,3] The use of electrically conductive fillers of polyconjugated polymers is more advisable. They give electrical conductivity to rubbers and improve the physical-mechanical indices due to the good compatibility of the components.

For the achievement of high electrical conductivity, the particles of electrically conductive fillers must form continuous electrically conductive structures. However, when using metal powders, the obtained polymer composite materials do not possess the necessary elastic properties. As a result, the rapid destruction of electrically conductive structures due to the deformation of products during operation occurs. Due to the good compatibility of the components, the use of electrically conductive additives of an organic nature (especially oligomeric and polymer compounds) allows for obtaining electrically conductive composition materials with good operational properties.[4–7]

The polyfunctional aromatic polyconjugated macromolecular compounds attract attention as electrically conductive additives. Such macromolecular compounds also possess thermal and radiation resistance, paramagnetism, semiconductivity, stabilizing and antistatic activity, solubility, meltability, and high reactivity in various chemical conversions.[5–10]

Taking into account this circumstance, we have synthesized the polyfunctional cooligomers of 1-naphthylamine with 1-naphthol, including reactive amine and hydroxyl groups, in elementary links. Cooligonaphthylamine naphthols (CONN) have been synthesized by the oxidative polycondensation of 1-naphthylamine with 1-naphthol in the presence of hydrogen peroxide.

8.2 EXPERIMENTAL METHODS AND MATERIALS

The synthesis of cooligomers of 1-naphthylamine and 1-naphthol was carried out in a three-necked flask with a volume of 250 mL, equipped with a thermometer, a mechanical stirrer, and a reflux condenser. About 7.20 g of naphthol, 7.15 g of 1-naphthylamine, 34 g of 30% solution of hydrogen peroxide (0.3 mol H_2O_2), and 85 mL of distilled water were placed in the flask. The reaction temperature was 363 K and the process duration was 4 h. The CONN samples were purified from the nonreacting part of the monomer by washing them with hot distilled water. The samples' drying was carried out at 373 K in a vacuum box.

The IR spectra of CONN were taken for thin films of 1-naphthylamine oligomers applied on monocrystals of NaCl on Specord M-80 spectrometer. The UV spectra of CONN samples were taken in ethanol using UV spectrometer-M-80.

The average values of M_w and M_n and also $\bar{M}_w / \bar{M}_n$ of the cooligomer samples were determined using a gel chromatography.[10] Three styrogel columns with porosity of 200, 500, and 1000 Å were used. The eluent—tetrahydrofuran calibration dependence was constructed, which is described by the equation $V_R = 30.2–3.8 \cdot \lg M$, where V_R is the retention volume, mL, and M is the molecular weight. The average molecular weights were calculated using the following formulas:

$$\bar{M}_w = \sum W_i M_i \tag{8.1}$$

$$\bar{M}_n = \frac{1}{\sum W_i / M_i}, \tag{8.2}$$

where W_1 is the mass part of fractions with molecular weight M_i (it was determined as a ratio of the area of the *i*th part of the chromatogram to the whole area).

The rubber mixture on the basis of butyl rubber (BR) and synthesized CONN was made by means of laboratory rollers. The rubber was plastified first on the rollers, and then the necessary ingredients, including CONN. Depending on the peculiarities of the mixture components, the temperature of the rollers was regulated in the front roller in the range of 305–315 K, and in the end roller in the range of 345–348 K.

Keeping the sequence of introduction of the components, the preparation of the rubber mixture was carried out as follows: (1) BR—0; stearine tech.—3 min; captax—5 min; thiuram—7 min; zinc oxide + oligomer—10 min; sulfur—15 min; total mixing time—20 min; and vulcanization temperature—313–323 K. Further, by means of hydraulic press PH-63, the obtained mixtures were pressed in special press molds as plates with a thickness of 1.5–2.0 mm. With the use of the standard knives, the samples of the necessary shapes and sizes for determination of physical-mechanical (tensile strength, specific elongation, residual deformation), electrical, and other properties were cut from the obtained plates.

The values of the physical-mechanical indices and the specific volumetric electrical conductivity (σ_v), $Ohm^{-1} \cdot cm^{-1}$ were determined and calculated on the standard methods and formulas.

8.3 RESULTS AND DISCUSSION

By the oxidative copolycondensation reaction of 1-naphthol with 1-naphthylamine in the presence of H_2O_2, CONN samples were synthesized. It has been shown that 1-naphthylamine and 1-naphthol in the presence of an aqueous solution of H_2O_2 at a temperature range of 343–368 K undergo the oxidative copolycondensation reaction and form the oligomer products with a yield ~23–88% (on monomers), depending on the synthesis condition.

It can be seen from Table 8.1 that with the temperature growth from 343 to 368 K, the yield of oligomer products increased from ~29% to 86%. The growth of the reaction duration up to 5 h and the ratio of H_2O_2: (NA +Nph) in the range of 1.0–3.0 (mol) also shows a noticeable positive influence on the yield of purposeful reaction products. According to the presented results, the optimal condition of carrying out the process is as follows: $[Nph]_o = [NA]_o = 0.75$ mol/L, $[H_2O_2]_o = 3.0$ mol/L, $T = 363$ K, $\tau = 4$ h. Under optimal regime, the yield of the oligomer products reaches 88.3%.

TABLE 8.1 Conditions and Yields of Oligomeric Products of the Oxidative Copolycondensation Reaction of 1-Naphthylamine with 1-Naphthol in the Presence of H_2O_2.

$[Nph]_0$, mol /L	$[NA]_o$, mol /L	$[H_2O_2]_o$, mol/L	T, K	τ, h	Cooligomer yield, %
0.75	0.75	1.5	343	4	29.4
0.75	0.75	1.5	353	4	58.6
0.75	0.75	1.5	363	4	74.2
0.75	0.75	1.5	368	4	86.1
0.75	0.75	1.5	363	1	30.3
0.75	0.75	1.5	363	2	54.7
0.75	0.75	1.5	363	3	67.8
0.75	0.75	1.5	363	5	75.2
0.75	0.75	1.5	363	8	78.6
0.50	0.50	1.5	363	4	82.4
0.50	0.50	2.0	363	4	86.7
0.50	0.50	3.0	363	4	88.3

The results of the elemental analysis of the samples and the values of the molecular weight indices of CONN samples synthesized in the various conditions, especially with changes in process temperature and the oxidizer content, show that these indices are noticeably different (Table 8.2.)

TABLE 8.2 Some Indices of COHH Samples Synthesized in the Various Conditions ($\tau = 5$ h).

№	$[NA]_o = [Nph]_o$, mol/L	$[H_2O_2]_o$, mol/L	T, K	Average molecular weight	
				$\bar{M}_w$	$\bar{M}_n$
1	0.5	1.5	343	645	330
2	0.5	1.5	353	660	375
3	0.5	2.0	363	730	395
4	0.5	3.0	363	810	580

In the IR spectra of CONN samples, the absorption bands of aromatic amine and hydroxyl groups, naphthalene ring, and isolated aromatic CH groups have been fixed (Figure 8.1).

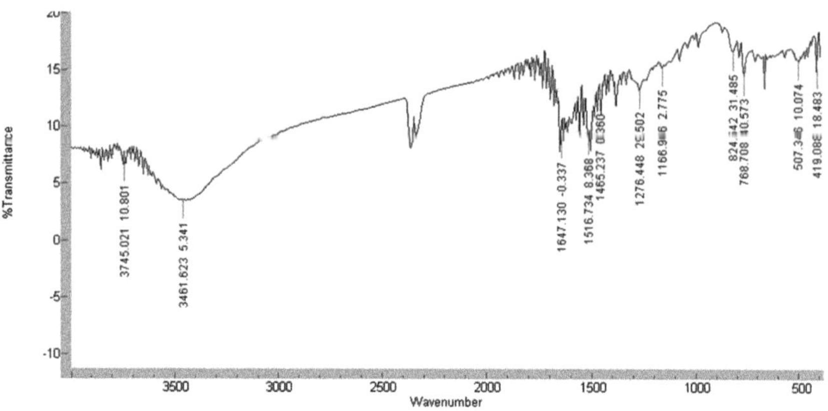

FIGURE 8.1 IR spectra of cooligonaphtylamine naphthol.

A characteristic peculiarity of the IR spectra of the cooligomers obtained by oxidative copolycondensation of NA+Nph system in the presence of H_2O_2 is that the absorption bands corresponding to the valence vibrations of NH, NH_2, and OH groups are overlapped and are fixed as one wide and intense peak (3100–3760 cm^{-1}). The absorption bands at frequencies of 1647, 1518, and 1465 cm^1 can be attributed to skeletal vibrations of the naphthalene ring. The absorption bands corresponding to aromatic CH groups in the naphthalene ring are observed in the form of numerous low-intensity maxima fixed in the range of 769–825 sm^{-1}.

Thus, based on the results of elemental and IR spectroscopic analyses, it can be concluded that the macromolecules of synthesized CONN samples consist mainly of the following structural units:

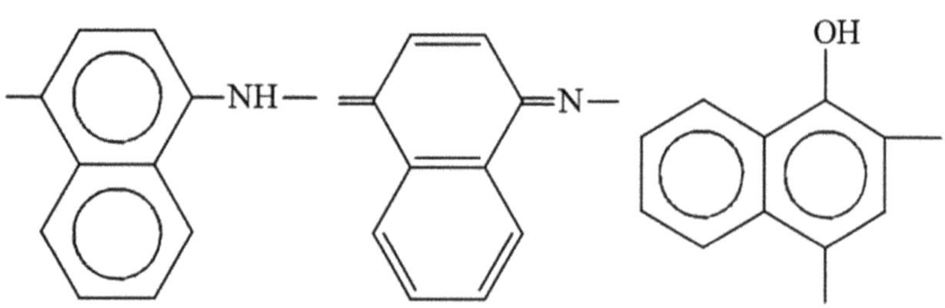

It is known from the scientific literature that the aniline oligomers consisted of phenylamine and quinoimine links

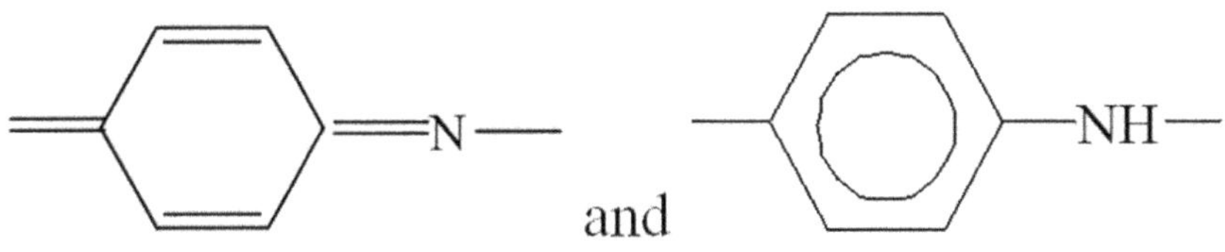

Similarly, the naphthylamine links in the macromolecules of CONN can exist in amine and quinoimine structures.

The UV spectra of the synthesized CONN samples confirm the proposed structure—a wide intensive peak of 210–230 nm with a maximum of 220 nm, and also less intensive absorption bands with maxima at 290 and 330 nm have been fixed. The first of them is the E-band ($\pi \rightarrow \pi^*$ excitation, and the second characterizes the $n \rightarrow \pi^*$ transitions of the undivided electrons of the nitrogen atom. The last peak indicates the availability of a system of polyconjugated bonds).

The synthesized CONN samples are black powders and are well-dissolved in polar organic solvents. They are melted under load at 373–398 K, depending on the synthesis conditions. Their composition and structure have been established by the methods of elemental, chemical, and IR spectral analysis, and molecular-weight indices—by gel-penetrating chromatography (Table 8.1).

CONN possesses the paramagnetism and electrical conductivity. The EPR spectra of solid powdery CONN samples have a singlet form ($g = 2.0023 \pm 0.0002$, $\Delta H = 0512$ MTs), the concentrations of paramagnetic centers in their composition are $(1.5–4.2) \times 10^{17}$ spin/cm^3, depending on the synthesis condition. Thus, oligonaphthylamine can be attributed to high-resistance polymer semiconductors; the values of the specific volumetric electrical conductivity of the CONN samples are $\sigma = \sim 10^{-12}–10^{-13}$ Ohm$^{-1}\cdot$cm^{-1}.

Consequently, these oligomers are of interest as an electrically conductive additive, but they can also act as the function of an antioxidant in the structure of compositions and thereby increase their heat- and thermal stability, and also their effective exploitation.

The synthesized CONN samples have been used as an active electrically conductive additive for the preparation of the rubber mixtures on the basis of rubbers, for example, butyl rubber (BR). In this case, the rubber mixtures on the basis of BR have been made according to the standard recipe of ingredients; however, instead of carbon black (partially) CONN samples (from 20.0 to 45 mass p. per 100 mass p. of rubber) have been used (Table 8.3).

TABLE 8.3 Some Parameters of the Obtained Vulcanizates Based on Butyl Rubber and CONN.

№	Ingredients	Mixtures (mass, g)			
		1	2	3	4
1	Butyl rubber	100	100	100	100
2	Technical stearine	3.0	3.0	3.0	3.0
3	Captax	0.65	0.65	0.65	0.65
4	DPhH	1.3	1.3	1.3	1.3
5	Zinc oxide	5.0	5.0	5.0	5.0
6	Carbon black	50	30	15	5
7	CONN	0	20	35	45
8	Sulfur	2.0	2.0	2.0	2.0

Properties							
Mixtures	Vulcanization time, min.	Tensile strength, MPa	Specific elongation, %	Residual deformation, %	Modulus of elongation, MPa		$\sigma \cdot 10^8$, $Ohm^{-1} \cdot cm^{-1}$
					200%	300%	
1.	40	19.5	710.0	4.2	9.4	14.1	
	60	18.5	700.0	4.5	9.0	13.2	
	80	19.4	663.3	4.3	9.8	14.3	
2.	40	22.4	675.6	7.9	8.8	12.2	0.29
	60	25.6	667.2	7.4	8.2	11.7	0.45
	80	25.3	596.4	6.8	8.6	12.8	0.51
3.	40	22.6	685.2	9.8	6.5	10.7	4.2
	60	22.9	654.0	11.8	6.9	11.1	6.9
	80	24.0	635.5	12.7	7.3	12.2	9.5
4.	40	26.0	674.2	11.8	6.2	8.8	35
	60	27.6	655.4	13.2	6.5	9.5	52
	80	28.9	589.3	14.8	6.9	10.6	68

Along with this, the thermal stability and lifetime of the obtained rubbers are increased, which is connected with the structural peculiarity of CONN; the condensed aromatic links in the aromatic polyconjunction chain stipulate high thermal stability and the availability of amine and hydroxyl groups in naphthalene rings—antioxidant activity. Since CONN samples exhibit high electrical conductivity, their joint use with electrically conductive carbon black allows to obtain the rubbers with a specific volumetric conductivity of 10^{-8}–10^{-6}Ohm^{-1}·cm^{-1}. The growth of CONN content from 20 to 45.0 m.p. (from the rubber mass) instead of carbon black leads to an increase in the specific electrical conductivity of the obtained rubbers. The percolation effect is reached at a content of −23.5 m.p of CONN for rubbers obtained from BR.

The reinforcing properties of CONN in the composition of the rubber composition have been probably stipulated with the optimal combination of such indices as small particle size, low density, and good compatibility of components of compositions.

8.4 CONCLUSION

By the oxidative copolycondensation reaction of 1-naphthylamine and 1-naphthol in the presence of H_2O_2, the polyfunctional polyconjugated soluble and meltable cooligomers, including the corresponding naphthylamine and naphthol links have been obtained.

The synthesized CONN samples demonstrate the paramagnetic and semiconductor properties. The EPR spectra of solid powdery CONN samples have a singlet form (g = 2.0023 ± 0.0002, ΔH=0,512 MTs), and the concentrations of paramagnetic centers in their composition are (1.5–4.2) × 10^{17} spin/cm^3. CONN can be attributed to high-resistance polymer semiconductors. The values of the specific volumetric electrical conductivity of the CONN samples are σ = ~ 10^{-12}–10^{-13} Ohm^{-1}·cm^{-1}, depending on the synthesis conditions.

It has been shown that with the partial substitution of carbon black of CONN in the composition of vulcanizate from BR, the obtained rubbers possess high operational indices (heat-physical, physical-mechanical, and electrical).

KEYWORDS

- **naphthylamines**
- **naphthols**
- **polyconjugated amines**
- **cooligonaphtylamine naphthol**

REFERENCES

1. Naarmann, H. *"Polymers, Electrically Conducting" in Ullman's Encyclopedia of Industrial Chemistry*; Wiley-VCH, 2002; p 429.
2. Kaverinskiy, V. S., Smekhov, F. M. *Electrical Properties of Paint and Varnish Materials and Coatings*; Khimiya Publishing House, 1990; p 158.
3. Valipour, A. Y.; Modhaddam, P. N.; Mamedov, B. A. Some Aspects of Chemical Procedures & Application Trends of Polyaniline as an Intrinsically Conductive Polymer. *Life Sci. J.* **2012,** *9,* 409–421.
4. Valipour, A. Y.; Modhaddam, P. N.; Mamedov, B. A. Synthesis of Hydrophilic Template as A Matrix Polymer for Preparationof Soluble Conductive Polyaniline Composites. *Arch. Sci. J.* **2012,** *65,* 14–20.
5. Pud, A.; Ogurtsov, N.; Korzhenko, A.; Shapoval, Q. Some Aspects of Preparation Methods and Properties of Polyaniline Blendsand Composites with Organic Polymers. *Prog. Polym. Sci.* **2003,** *28,* 1701–1753.
6. Ragimov, A. V.; Mamedov, B. A.; Gasanova, S. G. New Efficient Dielectric and Antistatic Materials Based on Oligoaminophenols. *Polym. Int.* **1997,** *43,* 343–348.
7. Mammadov, B. A.; Ahmadova, R. A.; Mashayeva, S. S.; et al. Preparation of Oligo-4-Aminophenol and ElectroconductiveComposition Materials on Its Basis. *Eur. Appl. Sci.* **2014,** *2,* 147–153.
8. Tefera, M.; Geto, A.; Tessema, M.; Admassie, S.; et al.. Simultaneous Determination of Caffeine and Paracetamol by Square WaveVoltammetry at Poly(4-Amino-3-Hydroxynaphthalene Sulfonic Acid)-Modified Glassy Carbon Electrode. *Food Chem.* **2016,** *210,* 156–162.
9. Bhandari, H.; Srivastav, R.; Choudhary, V.; Dhawan, S. K. Nhancement of Corrosion Protection Efficiency of Iron byPoly(Aniline-Co-Amino-Naphthol-Sulphonic Acid) Nanowires Coating In Highly Acidic Medium. *J. Thin Solid Films* **2010,** *519,* 1031–1039.
10. Doğan, F.; Şirin, K.; Kolcu, F.; Kaya, İ. Conducting Polymer Composites Based On LDPE Doped With Poly(AminonaphtholSulfonic Acid). *J. Electrostat.* **2018,** *94,* 85–93.

CHAPTER 9

Enlarging the Library of Degradable Triazole "Click" Polymers for Potential Biomedical Applications

TENGIZ KANTARIA[1], TEMUR KANTARIA[1], DAVID MAKHARADZE[2], NINO ZAVRADASHVILI[1], NODAR SULASHVILI[3], and RAMAZ KATSARAVA[1]

[1]*Institute of Chemistry and Molecular Engineering, Agricultural University of Georgia, Tbilisi, Georgia*

[2]*Departament d'Enginyeria Química, Universitat Politècnica de Catalunya, Barcelona, Spain*

[3]*Scientific Research-Skills Center, Tbilisi State Medical University, Tbilisi, Georgia*
Faculty of Medicine, Sulkhan-Saba Orbeliani University, Tbilisi, Georgia

ABSTRACT

A series of novel linear AA-BB-type poly(ester ether amide)s (PEEAs) and poly(ester ether)s (PEEs) containing 1,2,3-triazole moieties in the main chain were successfully obtained. The synthesis of the "click" polymers was performed via Cu(I)-catalyzed click step-growth polymerization (SGP) employing the efficient one-pot/two-step synthetic approach that we had previously developed. The confirmation of the polymer structure was achieved through the use of FT-IR and NMR spectroscopy. The newly synthesized materials were extensively characterized, including assessments of their yield, solubility in common organic solvents, film-forming

Advances in Engineering Materials: Preparation for Sustainable Process Development. Tamara Tatrishvili and A. K. Haghi (Eds.)

abilities, molecular weights, and their distribution. The obtained PEEAs and PEEs exhibited a quite satisfactory degree of polymerization, good film-forming properties, and favorable solubility behavior. The reported materials enlarge a library of available degradable triazole click polymers promising for a wide range of biomedical applications. The results validate once again the applicability of our innovative synthetic strategy for designing various classes of triazole-backbone degradable click polymers.

9.1 INTRODUCTION

The copper(I) catalyzed azide-alkyne cycloaddition (CuAAC) stands as a prominent tool within the realm of "click" reactions, showcasing its widespread utility across diverse domains in contemporary polymer and material science. This reaction has gained significant recognition due to its exceptional attributes, including rapid reaction kinetics, high efficiency, precise regioselectivity, tolerance toward diverse functional groups, and mild reaction conditions. Consequently, CuAAC click reactions have found extensive application in the synthesis of various polymer architectures, such as block-copolymers, cyclic polymers, dendrimers, and more complex branched structures such as cross-linked, star-shaped, and hyper-branched networks.[1–11] Furthermore, this technique has been instrumental in functionalizing and modifying the surfaces of polymeric materials.

Intrigued by its potential, synthetic polymer chemists have sought to leverage CuAAC for the synthesis of polytriazoles through step-growth polymerization (SGP) since 2004.[12–15] Despite notable advancements, this field remains in a state of continuous development, beset by numerous challenges and obstacles.[14–15] Specifically, there is a paucity of reports detailing the synthesis of aliphatic main-chain triazole degradable polymers through CuAAC-based click SGP. Noteworthy contributions include the pioneering work of Nagao and Takasu,[16] who successfully synthesized degradable polyesters (PEs), and the efforts of Galbis et al.,[17] who produced triazole PEs based on erythritol, bearing free hydroxyl groups. Beyond these achievements, click polymerization has also been harnessed for the fabrication of various other biomedical materials.[18–24]

It should be noted that click polymers with 1,2,3-triazoles in the main chain have some important advantages. To commence with, the triazole rings are known to be low-toxic moieties[25,26] with a high level of mimicry of the amide bond of native peptides: the triazole fragment is structurally similar to the amide bond of proteins in terms of configuration, distance, and planarity.[27] So, the incorporation of the triazole units in the polymeric chains is anticipated to improve the biocompatibility of the polymers. Furthermore, the insertion of rigid triazole fragments in the polymeric structure can notably improve the thermal characteristics of the polymers.[28] It is known that triazole's derivatives demonstrate a wide range of biological activities, including anti-inflammatory, antioxidant, antiviral, antimicrobial, antifungal, anticancer, analgesic, and so on.[29,30] Therefore, the click polymers with triazole rings in the backbone are expected to reveal some biological activities, and such polymers can be promising for creating new bioactive materials. It is worth noting that 1,2,3-triazole derivatives are intensively utilized in the design of various high-performance organic coatings such as antimicrobial, antifouling, anticorrosive, self-healing, and so on.[31] Thus, the triazole-enriched click polymers can be considered as perspective materials for creating effective polymeric coatings as well.

Moreover, an important advantage of click polymers lies in the quaternizable nature of 1,2,3-triazole rings, as documented in prior studies.[32,33] This characteristic facilitates subsequent postpolymerization functionalization of these polymers. The process of quaternization, involving the modification of triazole rings with diverse quaternizing agents such as halo-alkyls and polyethylene glycol (PEG) derivatives, presents opportunities for creating cationic systems featuring 1,2,3-triazolium moieties within the polymer backbone. This includes the development of cationic polymers and amphiphilic surfactants (e.g., when employing PEG-bromides as quaternizing agents) and the formation of cross-linked hydrogels (when utilizing bifunctional quaternizing agents like dibromo-PEGs, for instance). These approaches hold significant promise for a multitude of biomedical applications.

To synthesize new high-molecular-weight 1,2,3-trialoze containing degradable polymers, we have previously elaborated a new synthetic strategy of CuAAC-based click SGP.[28] Our innovative approach could be considered as more versatile compared to existing strategies and allows to synthesize various types (AB or AA-BB) and classes (PEs, poly(ester amide)s, poly(ester urea)s, their co-polymers, etc.) of aliphatic heterochain click polymers, avoiding the utilization of potentially explosive organic azides through safe and facile one-pot/two-step procedure. In the mentioned work,[28] we successfully synthesized a series of novel linear

AA-BB type 1,2,3-triazole rings-containing click PEs following the new strategy. Obtained on the basis of non-toxic starting materials (dicarboxylic acids and fatty diols), the new polymers possessed high molecular weights, improved thermal characteristics, good solubility in organic solvents, and film- and nanoparticles-forming properties. Very recently, we demonstrated the suitability of the new synthetic approach for designing novel hydrolytically degradable AB-type click PEs and Co-PEs as well.[34]

In this work, we report on the successful synthesis of two other classes of new linear AA-BB-type degradable click polymers via our efficient synthetic strategy: poly(ester ether amide)s (PEEAs) and poly(ester ether)s (PEEs). The key homo-bifunctional diyne monomers were synthesized by the interaction of sebacoyl chloride with propargyl alcohol or propargylamine according to Figure 9.1.

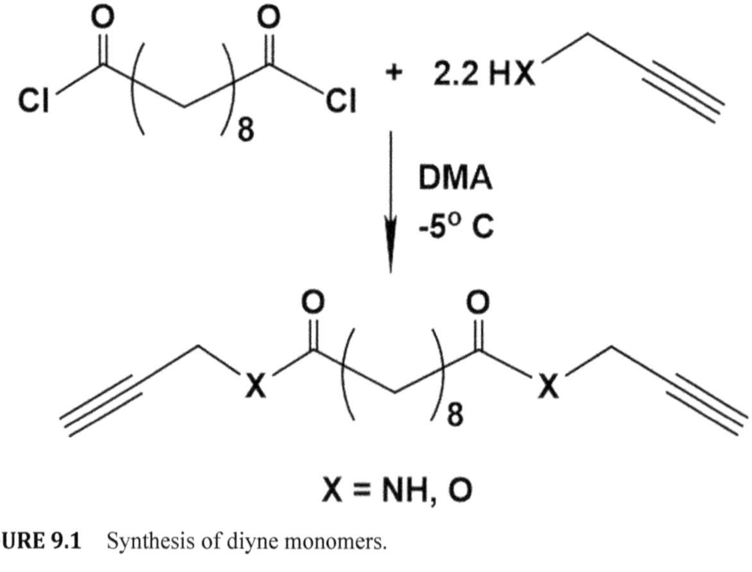

FIGURE 9.1 Synthesis of diyne monomers.

The synthetic precursors of target diazide monomers (bis-bromoacetyl derivatives) were obtained by the interaction of bromoacetyl bromide with di-, tri-, and tetra-ethylene glycol (Figure 9.2).

The synthesis of the new click polymers with triazole rings in the backbone was realized by one-pot/two-step SGP of the homo-bifunctional monomers in the presence of sodium azide employing copper(I) iodide/triethylamine (CuI/NEt_3) as a catalytic system (Figure 9.3).

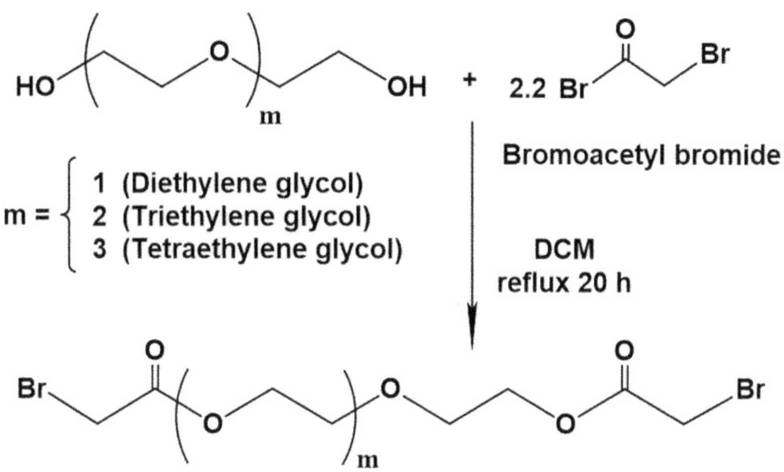

FIGURE 9.2 Synthesis of bis-bromoacetyl derivatives.

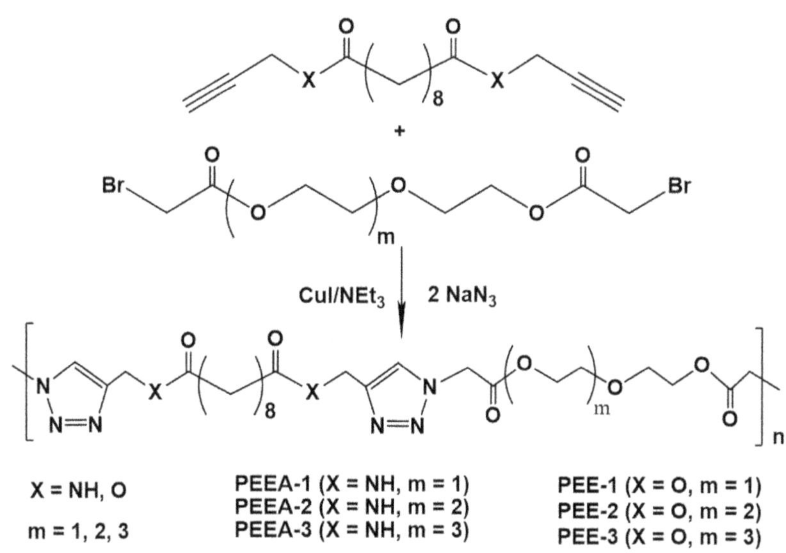

FIGURE 9.3 Synthesis of PEEAs (X=NH) and PEEs (X=O).

The confirmation of monomer and polymer structures was performed through the use of FT-IR and NMR spectroscopic techniques. Comprehensive polymer characterization encompassed the assessment of yield, solubility in organic solvents, film-forming attributes, determination of molecular weights, and evaluation of molecular weight distribution. The introduction of novel AA-BB-type PEEAs and PEEs serves to enrich the existing repository of degradable click polymers, broadening the spectrum of materials available for a diverse array of biomedical applications.

9.2 EXPERIMENTAL METHODS AND MATERIALS

9.2.1 MATERIALS

Broboacetyl bromide, propargyl alcohol, propargylamine, sodium azide, anhydrous sodium sulfate, copper(I) iodide, lithium bromide, aluminum oxide (alumina) (activated, basic, Brockmann I), acetone, ethanol, methanol, hexane, ethyl acetate, dichloromethane (DCM), N,N-Dimethylformamide (DMF), N,N-dimetyl-acetamide (DMA), and 1,1,1,3,3,3-hexa¬fluoro¬isopr opanol (HFIP) were purchased from Sigma-Aldrich (Germany). N-methyl-pyrrolidone (NMP) was obtained from Carl Roth (Germany) while triethyl-amine (TEA) was received from Lancaster (UK). All the starting materials, chemicals, and solvents purchased from commercial sources were exploited without additional purification steps.

9.2.2 CHARACTERIZATION

The polymers synthesized were characterized by Fourier-transform infrared spectroscopy (FT-IR), 1H nuclear magnetic resonance (NMR) spectroscopy, and gel permeation chromatography (GPC).

The IR analysis of samples was performed on the FT-IR spectrophotometer (Thermo Nicolet Avatar 370) using EZ OMNIC software. The wavelength range was 4000–400 cm^{-1}. The FT-IR spectra of monomers were recorded by exploiting a special Ge-based multibounce flat plate. To acquire the spectra of polymers, the films were prepared from HFIP solution, the solvent was evaporated at room temperature and the films were dried under reduced pressure to a constant weight.

The 1H and NMR spectra were recorded at 300 K on an NMR spectrometer (JEOL ECP 500) operating at 400 MHz. DMSO-d_6 was used as solvents and internal standards. The chemical shifts (δ) in the spectra are conventionally provided in parts per million (ppm).

The weight-average (M_w), number-average (M_n) molecular weights, and molecular weight distribution (Dispersity, *Đ*) of the polymers were determined on a GPC machine (Waters, USA) comprising two Styragel columns in DMF: HR4 and HR3 (7.8 mm × 300 mm) and coupled with a special high-pressure liquid chromatography pump (1525 Binary HPLC) and RI detector (2414, Waters). An eluent comprised of a 0.05 M solution of LiBr in DMF was employed. The injected volume of the sample was 100 μL, and the flow rate was 1.0 mL min^{-1}. The columns were calibrated with poly(methyl methacrylate) standards (M_n = 500–500,000 g mol^{-1}). Polymer sample solutions with a concentration of 5 mg mL–^{1} were prepared in the GPC eluent and filtered with PTFE filters (0.45 μm) before injection.

9.2.3 SYNTHESIS OF DIYNE MONOMERS

The key homo-bifunctional diyne monomers (di-propargyl amide/ester of sebacic acid) were synthesized by the interaction of sebacoyl chloride with propargylamine or propargyl alcohol according to Figure 9.1. The synthesis took place in a solution of DMA without the utilization of a tertiary amine as an HCl acceptor. In brief, propargylamine/propargyl alcohol (0.159 mol—a slight excess) was dissolved in 260 mL of DMA, the resulting solution was cooled down to –5°C, and 17.30 g of sebacoyl chloride (0.072 mol) was added dropwise under continuous stirring with a magnetic stirrer. The reaction proceeded at −5°C for 30 min and then at room temperature for a further 24 h. Subsequently, the solution was poured into water, and the container containing the resulting white precipitate was left overnight in a refrigerator. The solid product obtained was filtered, then dried under vacuum at room temperature, and recrystallized from the ethanol–hexane mixture.

Di-propargyl amide of sebacic acid. Yield: 68%. FTIR (cm^{-1}): 3280 (-NH-, amide, stretching vibr.), 2132 (C≡C), 1642 (–CO–, amide), 1552 (NH-, amide, bending vibr.). ^{1}H NMR (400 MHz, DMSO-d_6, δ): 1.28–1.32 (8H, m, C<u>H</u>$_2$); 1.54 (4H, m, C<u>H</u>$_2$CH$_2$CO); 2.16 (4H, t, C<u>H</u>$_2$CO); 3.04 (2H, t, sp-C<u>H</u>); 4.11 (4H, d, NHC<u>H</u>$_2$), 8.12 (2H, N<u>H</u>).

Di-propargyl ester of sebacic acid. Yield: 75 %. FTIR (cm^{-1}): 3238 (≡C–H), 2128 (C≡C), 1732 (–CO– ester), 1217 (C–O–C). ^{1}H NMR: (400 MHz, DMSO-d_6, δ): 1.29–1.35 (8H, m, C<u>H</u>$_2$); 1.59 (4H, m, C<u>H</u>$_2$CH$_2$CO); 2.30 (4H, t, C<u>H</u>$_2$CO); 2.93 (2H, t, sp-C<u>H</u>); 4.61 (4H, d, OC<u>H</u>$_2$).

9.2.4 SYNTHESIS OF BIS-BROMOACETYL DERIVATIVES (PRECURSORS OF DIAZIDE MONOMERS)

The synthetic precursors of target diazide monomers (bis-bromoacetyl derivatives) were synthesized by the interaction of bromoacetyl bromide with di-(DEG), tri-(TEG), and tetra-ethylene glycol (TTEG) according to Figure 9.2 above. The brief synthetic procedure is described here on the example of DEG-based bis-bromoacetyl derivative: to a solution of 6.09 g (0.057 mol) of DEG in 315 mL of DCM, 25.52 g (0.126 mol) of bromoacetyl bromide was added. The yellow solution obtained was subjected to reflux for 20 h, and the solvent was then removed under vacuum using a rotary evaporator, resulting in a viscous liquid with a brownish tint. The obtained crude product was dissolved in 40 mL of methanol and the solution was washed with 40 mL of hexane. The mixture was separated using a separated funnel and the procedure was repeated twice (i.e., the methanol solution of the crude product was washed three times with hexane). After the last separation of the mixture, the methanol solution of the monomer was dehydrated using anhydrous sodium sulfate, then it was filtered and the organic solvent was removed in a vacuum on a rotary evaporator, yielding a brownish viscous liquid. DEG- and TTEG-based bis-bromoacetyl derivatives were synthesized according to the same procedure.

DEG-based bis-bromoacetyl derivative. Yield: 69 %. FTIR (cm^{-1}): 1740 (–CO–, ester), 1282 (C–O–C, ester), 1123 (C–O–C, ether). ^1H NMR (400 MHz, DMSO-d_6, δ): 3.51 (4H, $C\underline{H}_2$–O–$C\underline{H}_2$), 4.11 (4H, COO–$C\underline{H}_2$), 4.21 (4H, CO–$C\underline{H}_2$).

TEG-based bis-bromoacetyl derivative. Yield: 76 %. FTIR (cm^{-1}): 1736 (–CO–, ester), 1278 (C–O–C, ester), 1115 (C–O–C, ether). ^1H NMR (400 MHz, DMSO-d_6, δ): 3.21–3.63 (8H, $C\underline{H}_2$–O–$C\underline{H}_2$–$C\underline{H}_2$–O–$C\underline{H}_2$), 4.12 (4H, COO–$C\underline{H}_2$), 4.22 (4H, CO–$C\underline{H}_2$).

TTEG-based bis-bromoacetyl derivative. Yield: 73 %. FTIR (cm^{-1}): 1744 (–CO–, ester), 1282 (C–O–C, ester), 1116 (C–O–C, ether). ^1H NMR (400 MHz, DMSO-d_6, δ): 3.13–3.47 (12H, $C\underline{H}_2$–O–$C\underline{H}_2$–$C\underline{H}_2$–O–$C\underline{H}_2$–$C\underline{H}_2$–O–$C\underline{H}_2$), 4.09 (4H, COO-$C\underline{H}_2$), 4.20 (4H, CO–$C\underline{H}_2$).

9.2.5 POLYMER SYNTHESIS

The target PEEAs and PEEs were synthesized according to the one-pot two-step synthetic strategy that we had previously developed[28,34] (Figure 9.3). The synthetic procedure is similar to the one applied for the AB-type click

PEs reported in our recent work.[34] The brief synthetic procedure for the click SGP of the PEEAs and PEEs is described here for PEEA-1 as an example: 1.902 g (5.46 mmol) of DEG-based bis-bromoacetyl derivative and 0.782 g (12.03 mmol) of sodium azide were dissolved in 8.35 mL of NMP and stirred for 3 h at room temperature. Subsequently, 1.511 g (5.46 mmol) of a di-propargyl amide of sebacic acid, 31.24 mg (0.16 mmol) of catalyst CuI (3 mol %), and 0.114 mL (0.82 mmol) of NEt_3 (ligand) were added to the reaction milieu. The reaction solution was stirred at room temperature for 24 h. Following the conclusion of the reaction, the solution was thinned by adding 8 mL of NMP, and the copper catalyst was removed by passing the crude polymer solution through a small column of activated basic alumina. Then, the obtained polymer solution was precipitated in water, filtered, washed with fresh portions of water, and dehydrated under vacuum at room temperature. Then, the polymer was twice precipitated from NMP solution (at a concentration of 10% w/v) to water, filtered, and dried under vacuum at room temperature to a constant weight. All the other polymers were synthesized employing the same synthetic procedure.

PEEA-1. Yield: 73 %. FTIR (cm^{-1}): 3297 (–NH–, amide, stretching vibr.), 3068 (C=C–H of the 1,2,3-triazole), 1752 (–CO–, ester), 1638 (–CO–, amide), 1544 (NH–, amide, bending vibr.), 1217 (C–O–C, ester), 1135 (C–O–C, ether). ^{1}H NMR (400 MHz, DMSO-d_6, δ): 1.01 (12H, CH_2– ($\underline{CH_2}$)$_6$–CH_2), 1.85 (4H, CO–$\underline{CH_2}$– (CH_2)$_6$–$\underline{CH_2}$–CO), 3.24–3.50 (4H, $\underline{CH_2}$–O–$\underline{CH_2}$), 4.02–4.11 (8H, COO–$\underline{CH_2}$ and NH–$\underline{CH_2}$–, stack), 5.17 (4H, CO–$\underline{CH_2}$–N), 7.68 (2H, –C=C$\underline{H}$ of triazole), 8.16 (2H, N$\underline{H}$).

PEEA-2. Yield: 94 %. FTIR (cm^{-1}): 3293 (–NH–, amide, stretching vibr.), 3063 (C=C–H of the 1,2,3-triazole moiety), 1744 (–CO–, ester), 1638 (–CO–, amide), 1544 (NH–, amide, bending vibr.), 1217 (C–O–C, ester), 1131 (C–O–C, ether). ^{1}H NMR (400 MHz, DMSO-d_6, δ): 1.07 (12H, CH_2–($\underline{CH_2}$)$_6$–CH_2), 1.82 (4H, CO–$\underline{CH_2}$–(CH_2)$_6$–$\underline{CH_2}$–CO), 3.23–3.65 (8H, $\underline{CH_2}$–O–$\underline{CH_2}$–$\underline{CH_2}$–O–$\underline{CH_2}$), 4.03–4.12 (8H, COO–$\underline{CH_2}$ and NH–$\underline{CH_2}$–, stack), 5.18 (4H, CO–$\underline{CH_2}$–N), 7.70 (2H, –C=C$\underline{H}$ of triazole), 8.14 (2H, N$\underline{H}$).

PEEA-3. Yield: 89 %. FTIR (cm^{-1}): 3297 (–NH–, amide, stretching vibr.), 3076 (C=C–H of the 1,2,3-triazole), 1752 (–CO–, ester), 1638 (–CO–, amide), 1543 (NH–, amide, bending vibr.), 1225 (C–O–C, ester), 1127 (C–O–C, ether). ^{1}H NMR (400 MHz, DMSO-d_6, δ): 0.99 (12H, CH_2–($\underline{CH_2}$)$_6$–CH_2), 1.84 (4H, CO–$\underline{CH_2}$–(CH_2)$_6$–$\underline{CH_2}$–CO), 3.14–3.45 (12H, $\underline{CH_2}$–O–$\underline{CH_2}$–$\underline{CH_2}$–O–$\underline{CH_2}$–$\underline{CH_2}$–O–$\underline{CH_2}$), 3.93–4.08 (8H, COO–$\underline{CH_2}$ and NH–$\underline{CH_2}$–, stack), 5.13 (4H, CO–$\underline{CH_2}$–N), 7.66 (2H, –C=C$\underline{H}$ of triazole), 8.07 (2H, N$\underline{H}$).

PEE-1. Yield: 93 %. FTIR (cm^{-1}): 3145 (C=C–H of the 1,2,3-triazole), 1744 (–CO–, ester), 1213 (C–O–C, ester), 1164 (C–O–C, ether). ^{1}H NMR (400 MHz, DMSO-d_6, δ): 0.87 (12H, CH_2–$(C\underline{H}_2)_6$–CH_2), 1.90 (4H, CO–$C\underline{H}_2$–$(CH_2)_6$–$C\underline{H}_2$–CO), 3.29 (4H, $C\underline{H}_2$–O–$C\underline{H}_2$), 3.91 (4H, COO–$C\underline{H}_2$-CH_2), 4.78 (4H, COO–$C\underline{H}_2$–C), 5.10 (4H, CO–$C\underline{H}_2$–N), 7.83 (2H, –C=C$\underline{H}$ of triazole).

PEE-2. Yield: 87 %. FTIR (cm^{-1}): 3141 (C=C–H of the 1,2,3-triazole), 1740 (–CO-, ester), 1213 (C–O–C, ester), 1168 (C–O–C, ether). ^{1}H NMR (400 MHz, DMSO-d_6, δ): 0.99 (12H, CH_2–$(C\underline{H}_2)_6$–CH_2), 2.07 (4H, CO–$C\underline{H}_2$–$(CH_2)_6$–$C\underline{H}_2$–CO), 3.30 (8H, $C\underline{H}_2$–O–$C\underline{H}_2$–$C\underline{H}_2$–O–$C\underline{H}_2$), 4.02 (4H, COO–$C\underline{H}_2$–CH_2), 4.91 (4H, COO–$C\underline{H}_2$–C), 5.20 (4H, CO–$C\underline{H}_2$–N), 7.93 (2H, –C=C$\underline{H}$ of triazole).

PEE-3. Yield: 85 %. FTIR (cm^{-1}): 3145 (C=C–H of the 1,2,3-triazole), 1752 (–CO–, ester), 1209 (C–O–C, ester), 1164 (C–O–C, ether). ^{1}H NMR (400 MHz, DMSO-d_6, δ): 0.97 (12H, CH_2–$(C\underline{H}_2)_6$–CH_2), 2.04 (4H, CO–$C\underline{H}_2$–$(CH_2)_6$–$C\underline{H}_2$–CO), 3.26 (12H, $C\underline{H}_2$–O–$C\underline{H}_2$–$C\underline{H}_2$–O–$C\underline{H}_2$–$C\underline{H}_2$–O–$C\underline{H}_2$), 4.02 (4H, COO–$C\underline{H}_2$–CH_2), 4.88 (4H, COO–$C\underline{H}_2$–C), 5.17 (4H, CO–$C\underline{H}_2$–N), 7.91 (2H, –C=C$\underline{H}$ of triazole).

9.3 RESULTS AND DISCUSSION

9.3.1 SYNTHESIS OF DIYNE MONOMERS

The detailed procedure of the synthesis of key homo-bifunctional diyne monomers (di-propargyl amide/ester of sebacic acid) was discussed earlier. The AA-BB-type monomers were synthesized by the interaction of sebacoyl chloride with propargylamine or propargyl alcohol (Figure 9.1). The yields of the obtained di-propargyl amide of sebacic acid and di-propargyl ester of sebacic acid were 68 and 75%, respectively. The presumed structure of the diyne monomers was proved by FT-IR and ^{1}H NMR spectroscopies.

9.3.2 SYNTHESIS OF BIS-BROMOACETYL DERIVATIVES (PRECURSORS OF DIAZIDE MONOMERS)

The synthesis of bis-bromoacetyl derivatives—precursors of diazide monomers—was provided above (see section on materials and methods). Three bis-bromoacetates were obtained by the interaction of bromoacetyl bromide with di-(DEG), tri-(TEG), and tetra-ethylene glycol (TTEG) (see Figure 9.2). The yields of the obtained bis-bromoacetyl derivatives were in the range of 69–76%. Their presumed structure was proved by FTIR and ^{1}H NMR spectroscopies.

9.3.3 SYNTHESIS OF THE CLICK PEEAS AND PEES

Six new triazole click polymers (three PEEAs and three PEEs) were synthesized via one-pot/two-step synthetic strategy that we had previously developed.[28,34] The detailed procedure of the synthesis of target click polymers was discussed earlier (see Section 9.2).

The polymers were obtained with quite high yields (73–94%). The molecular weight characteristics of the new click polymers such as weight-average (M_w), number-average (M_n) molecular weights, and molecular weight distribution (*Đ*) were determined using the GPC system with PMMA standards. Due to the insolubility of the PEEA-1 in DMF, it was not characterized by the GPC technique. According to the obtained data, given in Table 9.1, the new click polymers possess average molecular weights (M_w is in the range of 14,500–34,800 g mol^{-1}) and display moderately wide molecular weight distribution (*Đ* = 1.47–1.60). The degree of polymerization of the polymers is quite satisfactory for step-growth polymers (the DP is in the range of 16–42).

TABLE 9.1 Characteristics of the Obtained Click PEEAs and PEEs.

Polymer	M_w (g mol^{-1})	M_n (g mol^{-1})	Đ	DP
PEEA-2	14,500	9300	1.55	16
PEEA-3	16,600	11,400	1.47	18
PEE-1	34,800	23,100	1.51	42
PEE-2	28,300	17,800	1.60	30
PEE-3	27,100	17,200	1.58	27

Đ—Dispersity (M_w/M_n); DP—degree of polymerization.

To assess the film-forming abilities of the new polymers, the thin films were cast from HFIP solutions on polytetrafluoroethylene dishes, the solvent was allowed to evaporate at room temperature, and the generated films were finally dried in a vacuum at room temperature to a constant weight. With the exception of PEEA-1 and PEE-1, which formed brittle films, all the other click polymers revealed good elastic film-forming properties. The poor elasticity of the films produced by DEG-based polymers PEEA-1 and PEE-1 can be attributed to the highest rigidity of their molecular structure stipulated by the shortness of their monomeric units (only two $-CH_2-CH_2-O-$ fragments per monomeric unit).

9.3.4 SPECTRAL STUDIES

The structure of the click PEEAs and PEEs was investigated using FT-IR and NMR spectroscopy. Both methods confirmed the presumed structure of the new click polymers. The FT-IR study proved the presence of all IR-detectable functional moieties of the polymers. All the anticipated absorption peaks were revealed: 3297–3293 cm^{-1} attributed to the stretching vibration of amide NH, 3145–3063 cm^{-1} corresponding to stretching vibration of C=C–H of the 1,2,3-triazole ring, 1752–1740 cm^{-1} ascribed to ester C=O, 1638–1636 cm^{-1} distinctive for amide CO, 1544–1543 cm^{-1} expected for the bending vibration of amide NH, 1225–1209 cm^{1} characteristic of ester C–O–C moiety, and 1168–1127 cm^{-1} typical for ether C–O–C. The detailed ^1H NMR studies of the PEEAs and PEEs demonstrated precise regioselectivity of Cu(I) catalyst-assisted SGP: single signals of 1,2,3-triazole protons (C=C–H) at 7.66–7.93 ppm in the ^1H NMR spectra of the polymers confirmed the presence of exclusively 1,4-disubstituted 1,2,3-triazole rings in the macromolecular chains.

9.3.5 SOLUBILITY OF THE CLICK PEEAS AND PEES

The solubility of the new click PEEAs and PEEs in ten common organic solvents (10.0 mg of polymer in 1.0 mL of a solvent) is presented in Table 9.2. All the new polymers were insoluble in conventional solvents such as DCM, ethanol, ethyl acetate, acetone, and acetonitrile even at elevated temperatures. Overall, PEEs revealed better solubility as compared to PEEAs. All the PEEs displayed good solubility in common polar organic solvents such as NMP, DMF, DMSO, HFIP, and DMA at room temperature. Expectedly, the best solubility among the PEEAs was demonstrated by TTEG-based PEEA-3 which was readily soluble in DMSO and HFIP at room temperature and in NMP, DMF, and DMA upon heating. The poorest solubility of the PEEA-1, which was soluble only in DMSO, NMP, and HFIP, may be attributed to the increased rigidity of the molecular architecture conditioned by the shortness of the monomeric unit (only two $–CH_2–CH_2–O–$ fragments per monomeric unit). The poor solubility behavior of PEEAs compared to PEEs can presumably be ascribed to the increased rigidness of the macromolecular backbone and stronger intermolecular forces stipulated by the robust amide bonds.

TABLE 9.2 Solubility of the Click PEEAs and PEEs (10 mg of Polymer in 1 mL of Solvent).

Polymer	DMSO	NMP	DMF	DMA	HFIP	DCM	EtOH	Acetone	CH_3CN	EtAc
PEEA-1	+t	+t	–	–	+	–	–	–	–	–
PEEA-2	+t	+t	+t	+t	+	–	–	–	–	–
PEEA-3	t	+t	+t	+t	+	–	–	–	–	–
PEE-1	+	+	+	+	+	–	–	–	–	–
PEE-2	+	+	+	+	+	–	–	–	–	–
PEE-3	+	+	+	+	+	–	–	–		–

+soluble at room temperature; +t soluble upon heating; – insoluble.

9.4 CONCLUSION

Three new linear AA-BB-type poly(ester ether amide)s (PEEAs) and three poly(ester ether)s (PEEs) containing 1,4-disubstituted 1,2,3-triazole rings in the main chain were successfully synthesized through Cu(I)-catalyzed click step-growth polymerization (SGP) using homo-bifunctional monomers following one-pot/two-step synthetic strategy we have previously elaborated. The presumed structure of the new polymers was confirmed by FT-IR and NMR techniques. The click polymers with average molecular weights (M_w = 14,500 – 34,800 g mol^{-1}) and moderately wide molecular weight distribution (1.47–1.60) were obtained in quite good yields (73–94%). The degree of polymerization of the materials was quite satisfactory for step-growth polymers (DP = 16–42), allowing most of them to form elastic films. The click polymers reported in this work expand a library of available degradable polymers with functionally interesting 1,2,3-triazole rings in the macromolecular main chain. The new polymers obtained can be considered as having a potential for various biomedical applications.

ACKNOWLEDGMENTS

This work was supported by Shota Rustaveli National Science Foundation of Georgia (SRNSFG) [YS-21-2522].

KEYWORDS

- **poly(ester ether amide)s (PEEAs)**
- **poly(ester ether)s (PEEs)**
- **copper(i) catalyzed azide-alkyne cycloaddition (CuAAC)**
- **step-growth polymerization (SGP)**
- **1,2,3-triazole ring**

REFERENCES

1. Fournier, D.; Hoogenboom, R.; Schubert, U. S. Clicking Polymers: A Straightforward Approach to Novel Macromolecular Architectures. *Chem. Soc. Rev.* **2007,** 36, 1369–1380.
2. Binder, W. H.; Sachsenhofer, R. "Click" Chemistry in Polymer and Materials Science. *Macromol. Rapid Commun.* **2007,** *28,* 15–54.
3. Johnson, J. A.; Finn, M. G.; Koberstein, J. T.; Turro, N. J. Construction of Linear Polymers, Dendrimers, Networks, and other Polymeric Architectures by Copper-Catalyzed azide-alkyne Cycloaddition "click" Chemistry. *Macromol. Rapid Commun.* **2008,** *29,* 1052–1072.
4. Billiet, L.; Fournier, D.; Du Prez, F. Step-growth Polymerization and "click" Chemistry: The Oldest Polymers Rejuvenated. *Polymer* **2009,** *50,* 3877–3886.
5. Iha, R. K.; Wooley, K. L.; Nyström, A. M.; Burke, D. J.; Kade, M. J.; Hawker, C. J. Applications of Orthogonal "click" Chemistries in the Synthesis of Functional Soft Materials. *Chem. Rev.* **2009,** *109,* 5620–5686.
6. Liang, L.; Astruc, D. The Copper(I)-catalyzed alkyne-azide cycloaddition (CuAAC) "click" Reaction and its Applications. An Overview. *Coord. Chem. Rev.* **2011,** *255,* 2933–2945.
7. Altintas, O.; Vogt, A. P.; Barner-Kowollik, C.; Tunca, U. Constructing Star Polymers *via* Modular Ligation Strategies. *Polym. Chem.* **2012,** *3,* 34–45.
8. Tunca, U. Orthogonal Multiple click Reactions in Synthetic Polymer Chemistry. *J. Polym. Sci., Part A: Polym. Chem.* **2014,** *52,* 3147–3165.
9. Arseneault, M.; Wafer, C.; Morin, J.-F. Recent Advances in Click Chemistry Applied to Dendrimer Synthesis. *Molecules* **2015,** *20,* 9263–9294.

10. Geng, Z.; Shin, J. J.; Xi, Y.; Hawker, C. J. Click Chemistry Strategies for the Accelerated Synthesis of Functional Macromolecules. *J. Polym. Sci.* **2021,** *59,* 963–1042.
11. Huang, D.; Qin, A.; Tang, B. Z. Click Reactions in Polymer Synthesis. In *Encyclopedia of Polymer Science and Technology*; John Wiley & Sons, **2021**, pp 1–31.
12. Scheel, A. J.; Komber, H.; Voit, B. Novel Hyperbranched poly([1,2,3]-triazole)s Derived from AB_2 Monomers by a 1,3-dipolar Cycloaddition. *Macromol. Rapid Commun.* **2004,** *25,* 1175–1180.
13. Qin, A.; Lam, J. W. Y.; Tang, B. Z. Click Polymerization: Progresses, Challenges, and Opportunities. *Macromolecules* **2010,** *43,* 8693–8702.
14. Li, H.; Sun, J.; Qin, A.; Tang, B. Z. Azide-alkyne Click Polymerization: An Update. *Chinese J. Polym. Sci.* **2012,** *30,* 1–15.
15. Shi, Y.; Sun, J. Z.; Qin, A. Click polymerization: The Aurora of Polymer Synthetic Methodology. *Int. J. Polym. Sci., Part A: Polym. Chem.* **2017,** *55,* 616–621.
16. Nagao, Y.; Takasu, A.; "Click polyester": Synthesis of Polyesters Containing Triazole Units in the Main Chain via Safe and Rapid "click" Chemistry and their Properties. *J. Polym. Sci. Part A: Polym. Chem.* **2010,** *48,* 4207–4218.
17. Bueno, M.; Molina, I.; Galbis, J. A.; Degradable "click" Polyesters from Erythritol having free Hydroxyl Groups. *Polym. Degrad. Stabil.* **2012,** *97,* 1662-1670.
18. Liu, X.; Quan, L.; Tian, J.; Laquer, F.; Ciborowski, P.; Wang, D. The Syntheses of Click PEG – dexamethasone Conjugates for the Treatment of Rheumatoid Arthritis. *Biomacromolecules* **2010,** *11,* 2621-2628.
19. Wang, Y.; Zhang, R.; Xu, N.; Du, F.; Wang, Y.; Tan, Y.; Ji, S.; Liang, D.; Li, Z. Reduction-degradable Linear Cationic Polymers as Gene Carriers Prepared by Cu(I)-catalyzed azide - alkyne Cycloaddition. *Biomacromolecules* **2011,** *12,* 66-74.
20. Guo, J.; Wei, Y.; Zhou, D.; Cai, P.; Jing, X.; Chen, X.; Huang, Y. Chemosynthesis of poly(ε -lysine)-analogous Polymers by Microwave-assisted Click Polymerization. *Biomacromolecules* **2011,** *12,* 737-746.
21. Eissa, A.; Khosravi, E. Synthesis of a New Smart Temperature Responsive Glycopolymer *via* Click-polymerisation. *Eur. Polym. J.* **2011,** *47,* 61-69.
22. van Dijk, M.; Nollet, M. L.; Weijers, P.; Dechesne, A. C.; van Nostrum, C. F.; Hennink, W. E.; Rijkers, D. T. S.; Liskamp, R. M. J. Synthesis and Characterization of Biodegradable Peptide-based Polymers Prepared by Microwave-assisted Click Chemistry. *Biomacromolecules* **2008,** *9,* 2834-2843.
23. Abid, M.; Aden Ali, M.; Bernard, J.; Abid, S.; Fleury, E.; El Gharbi, R. Preparation of New poly(ester triazole) and Poly(amide triazole) by "click chemistry". *Polym. Sci. Ser. B* **2014,** *56,* 290–297.
24. Rivas, M. V.; Petroselli, G.; Erra-Balsells, R.; Varela, O.; Kolender, A. A. Synthesis, Characterization and Chemical Degradation of Poly(ester-triazole)s Derived from D-galactose. *RSC Adv.* **2019,** *9,* 9860–9869.
25. Sheehan, D. J.; Hitchcock, C. A.; Sibley, C. M. Current and Emerging azole Antifungal Agents. *Clin. Microbiol. Rev.* **1999,** *12,* 40–79.
26. Lima-Neto, R. G.; Cavalcante, N. N. M.; Srivastava, R. M.; Mendonça Jr, F. J. B.; Wanderley, A. G.; Neves, R. P.; dos Anjos, J. V. Synthesis of 1,2,3-triazole Derivatives and *in vitro* Antifungal Evaluation on *Candida* Strains. *Molecules* **2012,** *17,* 5882–5892.
27. Ahmad Fuaad, A. A. H.; Azmi, F.; Skwarczynski, M.; Toth, I. Peptide Conjugation *via* CuAAC "click" Chemistry. *Molecules* **2013,** *18,* 13148–13174.

28. Kantaria, Ten.; Kantaria, Tem.; Titvinidze, G.; Otinashvili, G.; Kupatadze, N.; Zavradashvili, N.; Tugushi, D.; Katsarava, R. New 1,2,3-triazole containing Polyesters via Click step-growth Polymerization and Nanoparticles made of them. *Int. J. Polym. Sci.* **2018,** *2018,* 6798258. https://doi.org/10.1155/2018/6798258
29. Singh, G.; Sharma, P.; Dadhwal, S.; Garg, P.; Sharma, S.; Mahalan, N.; Rawal, S. Triazoles Impinging the Bioactivities. *Int. J. Curr. Pharm. Res.* **2011,** *3,* 105–118.
30. Kharb, R.; Sharma, P. C.; Yar, M. S. Pharmacological Significance of Triazole Scaffold. *J. Enzyme Inhib. Med. Chem.* **2011,** *26,* 1–21.
31. Kantheti, S.; Narayan, R.; Raju, K. V. S. N. The Impact of 1,2,3-triazoles in the Design of Functional Coatings. *RSC Adv.* **2015,** *5,* 3687–3708.
32. Abdelhedi-Miladi, I.; Obadia, M. M.; Allaoua, I.; Serghei, A.; Romdhane, H. B.; Drockenmuller, E. 1,2,3-triazolium-based Poly(ionic liquid)s Obtained through Click Chemistry Polyaddition. *Macromol. Chem. Phys.* **2014,** *215,* 2229–2236.
33. Aizpurua, J. M.; Fratila, R. M.; Monasterio, Z.; Pérez-Esnaola, N.; Andreieff, E.; Irastorza, A.; Sagartzazu-Aizpurua, M. Triazolium Cations: From the "click" Pool to Multipurpose Applications. *New J. Chem.* **2014,** *38,* 474–480.
34. Kantaria, Ten.; Kantaria, Tem.; Vanishvili, A.; Kvinikadze, S.; Klinger, D.; Katsarava, R. Synthesis of New Degradable AB-type Polyesters with 1,2,3-triazole rings in the Backbone *via* "click" Step-growth Polymerization. *Asian J. Chem.* **2023,** *35,* 649–654.

CHAPTER 10

Chitosan-Based Composite Sponges with Konjak Glucomanan and Graphene for Potential Application as Hemostatic Materials

ZOIA HAHOLKINA, MAITE RICO, REBECA BOUZA, and BELEN MONTERO

Polymer Group, Physics Department, University of A Coruña, Polytechnic University School, Ferrol Campus, Ferrol, Spain

ABSTRACT

Porous materials based on chitosan (CS) with microparticles were obtained without the use of organic solvents for potential hemostatic application. CS-based particles could encourage and promote prolonged drug release to improve wound healing. According to IR spectroscopy results, the reaction between CS and sodium tripolyphosphate (TPP) was confirmed. The morphology and structure of sponge materials after the process of formation (by SEM) are cellular with CS particles. The resulting materials showed enough porosity (about 40%). In addition to the CS base, graphene (GF) and konjac glucomannan (KG) were used. Each of these components with its hemostatic mechanism contributes to the improvement of the properties of sponges. With the incorporation of KG and GF into the obtained sponges, based on CS, the water absorption rate for obtained materials has increased considerably (it is about 98% for composition CS/KG/GF).

Advances in Engineering Materials: Preparation for Sustainable Process Development. Tamara Tatrishvili and A. K. Haghi (Eds.)

10.1 INTRODUCTION

The study and application of materials for bleeding cessation are of great importance to different areas of medicine.[1] The most common cause of death in trauma is bleeding.[1,2] Therefore, new and improved hemostatic materials are developed every year.[2] Local hemostatic agents are drugs of external action, which have almost no effect on the coagulation and anticoagulation systems of blood. The variety of release forms provides the convenience of use in the field and fully meets the needs of emergency care for any bleeding. Present-day chitosan is considered to be the most effective substance for the development of hemostatic materials. It has been investigated[3–7] that the hemostatic mechanism of CS is based on erythrocyte adhesion, thrombocyte aggregation, and stimulation of components that promote blood clotting The properties of chitosan make it possible to produce different forms of materials (powders, sponges, dressings) with the absence of cytotoxicity, genetic toxicity, and without causing hemolysis, have sufficient biological compatibility and can biodegrade.[8,9] One of the most important properties of this type of material is its adsorption ability. A study by Jayakumar et al.[10] described a methodology to produce CS sponges that are capable of absorbing liquid about 20 times their dry mass. Based on CS with different molecular weights, membranes can also be produced that significantly improve tissue regeneration and promote hemostasis.[11] The most famous CS-based hemostatic materials are HemCon Bandage from the company HemCon (USA) and powder Celox developed by the company SAM (USA).[12] The range of production methods for most CS materials often consists of electrospinning, particle leaching, and gas-foaming processes by the use of foaming agents.[13–15] But using organic solvents and the special temperature conditions are significant disadvantages of these methods.

Ionic gelation is a facile and efficient technique to create composites based on CS with microparticles due to the use of not noxious chemicals, a simple process, and the possibility of controlling the size of particles.[16] CS composites can be synthesized by ionic gelation using counter-ions as cross-linking agents such as TPP due to its high complexability and ability to control particle size.[17,18] This method offers advantages of obtaining medical treatment. Since it can be carried out in nonconcentrated aqueous solutions, it can easily control reaction conditions and particle size and can be used to encapsulate drug molecules.[18]

The hemostatic mechanism of CS composites with GF based on electrostatic interaction with erythrocytes, accelerates red blood cell adhesion, platelet adhesion, and activation. Such hemostatic have strong adhesion, biocompatibility, biodegradability, nontoxicity, antimicrobial activity, and nonantigenicity.[7, 19–21]

Application of several substances (CS, KG, and GF) with potential hemostatic action at once may lead to synergy of their properties.

10.2 EXPERIMENTAL METHODS AND MATERIALS

10.2.1 FABRICATION OF CHITOSAN HEMOSTATIC SPONGES

Cross-linked CS-based sponges were obtained by ionic gelation method through the electrostatic interaction between a solution of CS at pH 4.5 (in acetic acid solution) and TPP at pH 4.5 (solution prepared in ultrapure water). To obtain the suspension, the TPP solution was added slowly to the 10 mL of CS solution, varying the TPP volume in the range of 0.5–10 mL (Figure 10.1). Synthesized samples with a ratio CS/TPP 5:1 were used. The mixture was kept under magnetic stirring for 40 min at room temperature. The colloidal suspensions were centrifuged. The final stage of obtained CS-based sponges was placed at −20°C with subsequent freeze-drying in a lyophilizer at a temperature of −80°C and 9.9 bar pressure.[15]

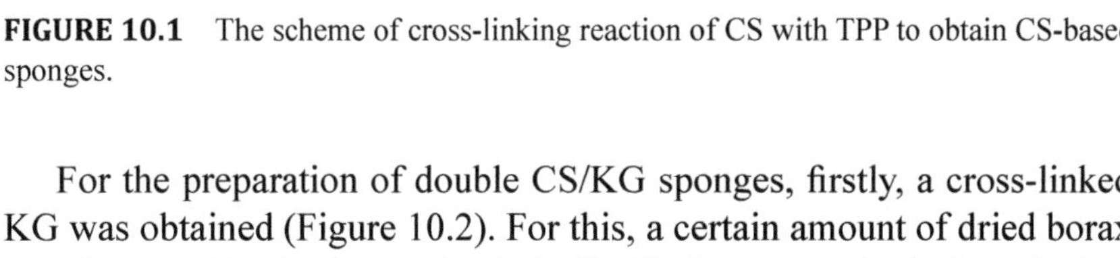

FIGURE 10.1 The scheme of cross-linking reaction of CS with TPP to obtain CS-based sponges.

For the preparation of double CS/KG sponges, firstly, a cross-linked KG was obtained (Figure 10.2). For this, a certain amount of dried borax powder was dissolved completely in distilled water to obtain the solution. Then, the KG (the ratio of KG to borax was 30:1) was dissolved in the borax solution under stirring at 80°C for 0.5 h. The resulting solution was added to cross-linked CS immediately after adding TPP. The mixture of CS and KG was kept under magnetic stirring for 40 min at room temperature.

To prepare the sponges with GF, first, a suspension was prepared by ultrasonication of the 1% wt. of GF in water for 2.5 min (Figure 10.3). Then, the prepared suspension was added to the crosslinked CS or cross-linked CS/KG suspension and ultrasonicated for another 2.5 min.

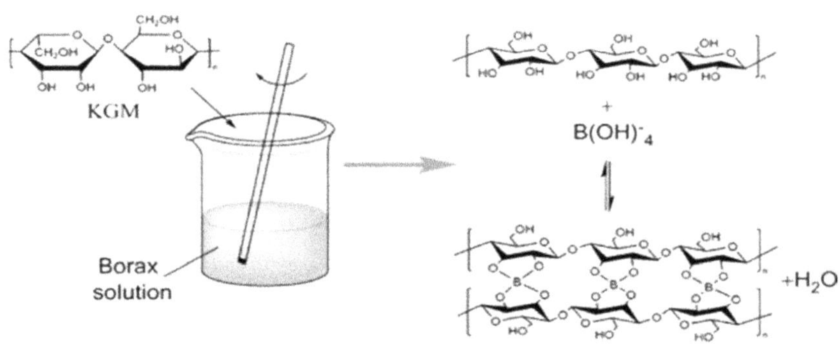

FIGURE 10.2 The scheme of cross-linking reaction of KG with borax to obtain KG-based sponges.

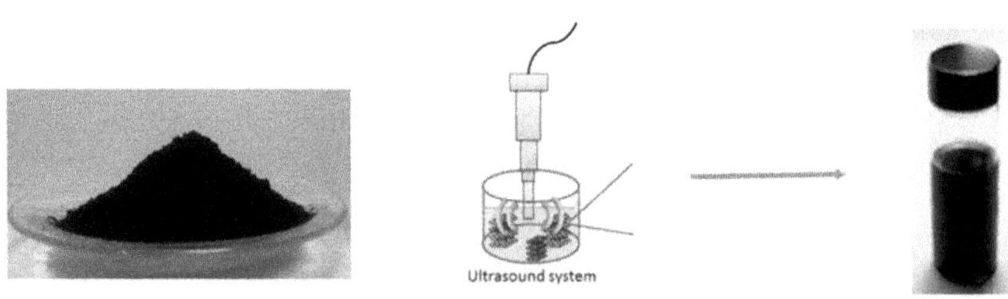

FIGURE 10.3 The scheme of preparation the graphene suspension.

10.2.2 OBSERVATION OF SURFACE MORPHOLOGY

The samples of sponges were fractured in liquid nitrogen and sprayed with gold plating for 180 s. The surface morphology of sponges was analyzed using scanning electron microscopy (SEM) equipment (JSM 7200F, Jeol, Japan). Particle size estimation was made using the ImageJ application.

10.2.3 FTIR ANALYSIS

The ATR-FTIR spectra were obtained on a Jasco 4700 spectrometer equipped with a MIRacle™ ZnSe Single Reflection Horizontal ATR accessory (Jasco, Japan). The tests were made with 100 scans between 4000 and 600 cm^{-1} with a resolution of 1.0 cm^{-1} and subjected to baseline correction.

10.2.4 POROSITY OF SPONGES

Porosity was determined by pycnometry using the device using an Accupyc 1340 II pycnometer (Micromeritics, Inc.). The gas used in the measurement is ultra-pure He (>99.9999%), Alphagaz I (supplied by Air Liquide).

Porosity (*Pt*) was calculated according to the following formula:

$$Pt = \frac{\mathrm{Vp}}{\mathrm{V}t} \tag{10.1}$$

where *Vp*—pore volume and *Vt*—total volume.

10.2.5 MECHANICAL PROPERTIES

The mechanical properties of obtained sponges including the compressive modulus were evaluated using Instron Universal Testing Machine 5565 (Instron, England) with a 100 N load cell and a crosshead speed of 2 mm min^{-1}. Cylindrical sponges with a diameter of 25 mm and a height of 10 mm were used for the compression tests. Compressive modulus was defined as the slope of the linear region of the strain/stress curves, corresponding to 5–15% strain. Data of the compressive strain correspond to the strain at the yield point. The results were reported as the mean ± standard error of at least five samples.

10.2.6 WATER ABSORPTION RATE

We studied the absorption rate of our materials according to the method described in Ref [7]. One liter of phosphate-buffered saline (137 mM NaCl, 2.7 mM KCl, and 10 mM with pH 7.4 at 25°C) was prepared.

The ion concentration in this solution is identical to the ion concentration in human serum or on the wound surface.[7]

Preliminary preparation of samples consisted of preparing sponges of the required size (0.5 × 0.5 cm) and drying and removing residual water to a constant weight. For this, sponges were put into a drying oven (80°C), weighed every 5 min until the variation was less than 0.0005 g, and then placed in a petri dish. After this, a model solution was added to the petri dish, the weight of which was 40 times greater than the weight of the sample. All the samples in petri dishes were placed in a drying oven and kept for 30 min at 37 (±1)°C. To remove excess solution from the surface of the sample, one end of it was clamped with tweezers for 30 s. To calculate the water absorption rate, the following formula was used (10.2):

$$\text{Water absorption rate} = \frac{m_2 - m_1}{m_1} \times 100\% \qquad (10.2)$$

m_1—sample weight before the adsorption of phosphate buffer solution (dried sample);

m_2—sample weight after the adsorption.

10.2.7 MATERIALS

The following materials were used: chitosan (CS) with a medium molecular weight purchased from Sigma-Aldrich (Spain); sodium tripolyposfat (TPP) (99.8% purity) by TermoFicher (Germany); acetic acid (99.8% purity) purchased from Cienytech; Konjak glucomanan by Biosynth (Carbosynth) (United Kingdom) (99% purity); and graphene (GF) (99% purity) purchased from the GreenTech (Spain).

10.3 RESULTS AND DISCUSSION

10.3.1 THE MORPHOLOGY OF THE SPONGES

Most often hemostatic sponges have a porous structure. It has not been confirmed whether a wide range of pore sizes is beneficial for hemostasis.[7] Figure 10.4 shows the general form of the CS-based sponge and the SEM (b-f) images of the sponges. The obtained sponges are characterized by the formation of a cellular structure. Numerous microparticles are also found, the size of which can vary from 0.5 to 10 μm. However, the results of this study showed that porous CS-based hemostatic sponges with a wide range of pore sizes can be produced by ionic gelation, and this will be useful for further investigation of CS-based hemostatic materials.

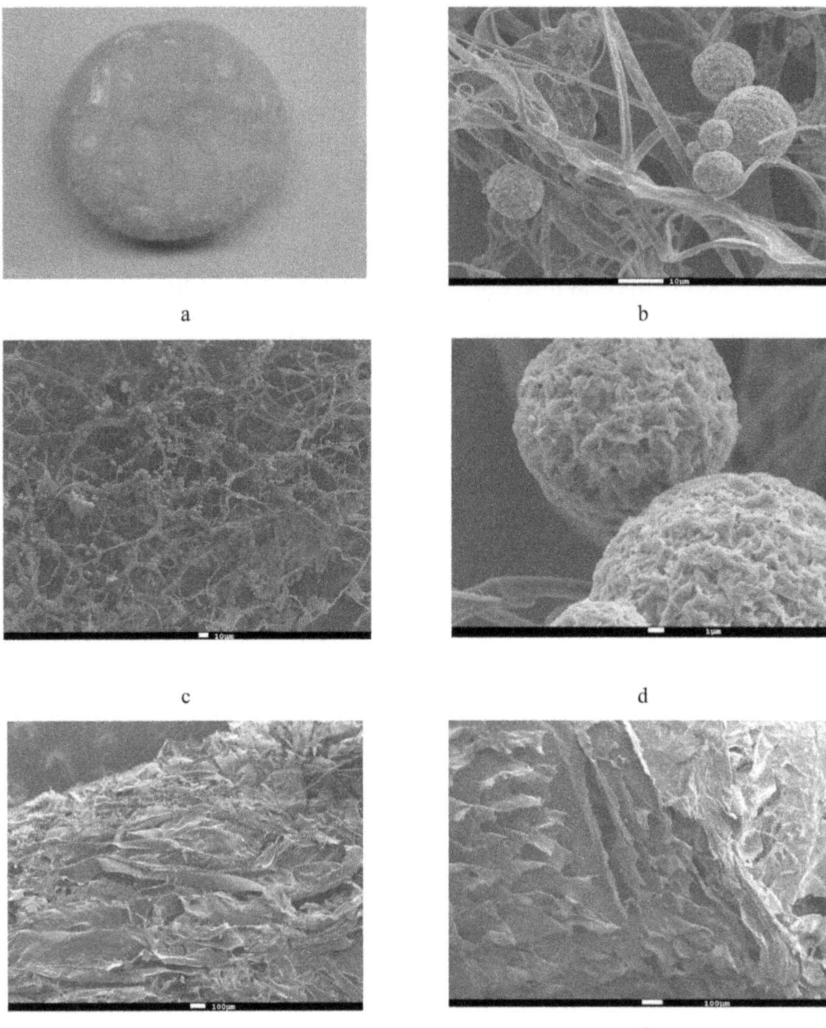

FIGURE 10.4 CS-sponge (a) and the SEM images of CS (b, c, d), KG (e), and CS/GF/KG sponges (f).

The formation of a cellular structure with different cell sizes could be attributed to the influence of the TPP content on the cross-linked network structure inside the hydrogel. The formation of cellular structures with different cell sizes can be explained by the influence of TPP concentration during the crosslinking process. Different ratios of CS and TPP (20:1; 10:1; 5:1, and 2:1) were investigated by Hoang et al.[16] The optimal structure (cellular with particles) in our opinion was observed for a CS/TPP ratio of 5:1. At low TPP content, no microparticles were observed, but at high crosslinking agent concentration, it was observed that the dense porous structure reduces the water accommodation space and swelling rate of the composite sponge, which may reduce the ability of the composite sponge to coagulate rapidly. The porous structure was not observed for KG samples. For samples CS/GF/KG, a layered structure with heterogeneous regions on the surface of the layers was observed. These heterogeneous regions are most likely chitosan microparticles incorporated into the material.

10.3.2 FTIR SPECTRA OF CS-BASED SPONGES

The FTIR spectra results, neat CS, and cross-linked CS obtained in the ratio CS/TPP 1:5 sponges of were shown in Figure 10.5. When studying the IR spectrum of chitosan, the following characteristic absorption signals, cm^{-1}, were observed: in the region 3353–3291 cm^{-1} due to valence vibrations of primary amino groups and –OH–groups attached to the pyranose ring; the signal in the region 2918–2872 cm^{-1} due to valence vibrations of the pyranose ring and $-CH_2OH$ attached to it, caused by valence vibrations of methylene groups. The most interesting from the point of view of estimating the chemical structure of chitosan is the region at 1500–1800 cm^{-1}. In Figure 10.5, there were three obvious absorption peaks on the spectral lines at 1699, 1643, and 1238 cm^{-1}, which corresponded to carbonyl stretching vibration, C–N bond stretching and NH bending vibration, respectively, suggesting that secondary amide groups existed in the cross-linked composite sponges. The absorption signals in the region 1649–1588.3 are due to valence vibrations of amide I and amide II bonds. Also, for chitosan, the band at 1093 cm^{-1} was attributed to C–O stretching.

Two absorption bands can be seen in the spectra of all samples: 1647–1661 cm^{-1}, which is responsible for the valence vibrations of the ν_{OH} bond, for the valence vibrations of the $\nu_{C=O}$ bond of the amide group, and an absorption band with a maximum at 1595–1599 cm^{-1}, corresponding to the deformation vibrations of the $\delta_{N\text{-}H}$ bond in amino groups.

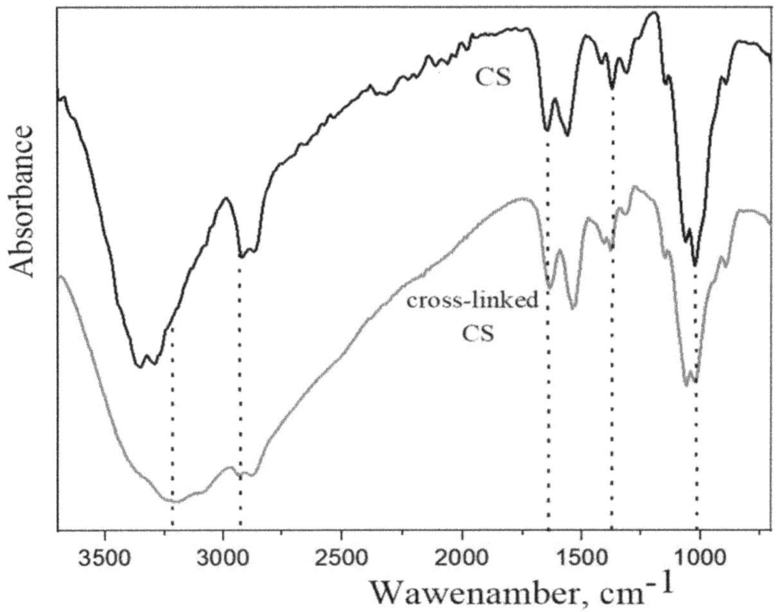

FIGURE 10.5 FTIR spectra results, neat CS, and cross-linked CS/TPP (1:5) sponge.

Redistribution and change in bandwidths may indicate the cross-linking reaction of CS with TPP.

10.3.3 POROSITY OF THE SPONGES

To evaluate the water retention capacity of nanocomposite sponges, we studied their porosity. It has been shown[22] that higher porosity of wound dressings promotes effective absorption of wound exudate and prevention of wound infections.

The important parameters related to the obtained materials, such as porosity, density, and maximum pore volume, are exhibited in Table 10.1

For example, for sponges CS/GF, increased porosity was observed compared to CS. At the same time, the maximum pore size decreased. This may be due to the peculiarities of graphene distribution in the material.

TABLE 10.1 Porosity, Density, and Maximum Pore Volume of the Obtained Materials.

Samples	Porosity (%)	Density (g/cm³)	Maximum pore volume (cm³/g)
CS	39,4	3,278	0,154
CS/GF	46,6	2,193	0,097
CS/KG	41,2	2,095	0,045
KG	37,5	3,412	0,153
CS/KG/GF	47,1	2,562	0,126

For CS/KG/GF, the porosity value is highest for the sponges analyzed. The sponge density value is also higher for CS/KG/GF. The sponge density value is higher for one-component systems, CS and KG, which we can explain by the compaction of the material during cross-linking of both chitosan and konjac glucomanan and the influence of graphene.

10.3.4 WATER ABSORPTION RATE

The ability of hemostatic sponges to absorb tissue fluid near a wound correlates with water absorption capacity. Sponges with a higher water absorption rate will absorb more blood at one time, which is important for an effective hemostasis.[7] Table 10.2 and Figure 10.6 show the water absorption rates of the different sponges in 30 min. According to the results, the incorporation of KG and GF into the CS sponges improved the water absorption rate compared with the net chitosan sponge. Previous studies[15, 25] have shown that the water absorption rate of the sponges is influenced by their structure and hydrophilicity.

TABLE 10.2 Water Absorption Rate of the Obtained Composites.

Samples	Water absorption rate, %
CS	94,3
CS/GF	95,4
CS/KG	97,5
KG	91,4
CS/KG/GF	98,1

For all the chitosan-based composites obtained, the water adsorption rate (%) increases with the addition of other KG and GF or these two components together.

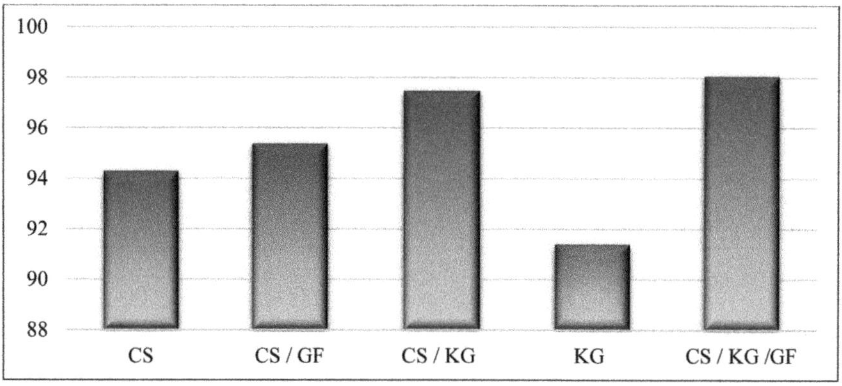

FIGURE 10.6 Water absorption rate (%) of the obtained composites.

We observed a definite correlation between the porosity and water adsorption rate of our materials. A high percentage of porosity was indicated for the samples CS/KG which correlates with the water absorption rate of 97.5%. These data correlate with higher porosity values. Sponges with three components recorded the highest absorption rates of water (98.1%). Probably, there is a synergistic effect of the water adsorption capacity of CS and KG, which individually have high values of this parameter.

10.3.5 MECHANICAL PROPERTIES

The hemostatic sponges must have sufficient mechanical strength to effectively reduce blood loss and be easily removed.[7] This is explained by the fact that the wound dressing is in close contact with the surface of the skin, it is necessary to ensure its sufficient mechanical strength to control the patient's comfort.[23] Thus, the compression stress–strain test was performed to assess the mechanical properties of the sponges. The values of the modulus of compressive strength of the sponges are presented in Table 10.3.

The obtained samples are characterized by rather low tensile modulus values. However, when these materials are obtained from chitosan and graphene, the modulus value increases noticeably from 0.03 to 0.4 MPa (for 13 times). For sponges CS/KG/GF, the values of modulus also raised significantly (7.6 times) compared to net CS-based sponge.

TABLE 10.3 Modulus of Compressive Strength.

Samples	Modulus, MPa
CS	0.03
CS/GF	0.4
KG	0.01
CS/KG	0.12
CS/KG/GF	0.23

10.4 CONCLUSION

A method for the preparation of chitosan sponges with KG and GF was developed. All materials were obtained by a combination of ionic gelation and freeze drying methods without organic solvents from non-toxic materials. The progress of chitosan cross-linking reaction was confirmed through the change in the character of FTIR absorption bands.

The prepared sponges showed noticeable porosity (about 40%) and better water absorption rate, which is favorable for the absorption of wound exudate; they can provide the right environment for optimal wound healing. According to an analysis of the SEM images, we can conclude that the structure of the obtained materials is cellular with microparticles. The presence of particles can significantly contribute to the delivery of wound-healing drugs and their prolonged release. The compressive mechanical tests showed that the modulus of compressive strength of the CS/KG/GF nanocomposite sponges was improved compared to net CS and KG sponges. The use of KG and GF as additives to the chitosan sponges contributes to a marked increase in the compressive modulus of the obtained CS/KG/GF materials compared to the single-component materials. The highest water adsorption rate results were also obtained for composites that included CS, KG, and GF at once. This can be attributed to the synergistic contribution of the components to the properties of these materials. To summarize, obtained sponges, especially CS/KG/GF, have great potential for wound dressings application.

ACKNOWLEDGMENTS

The authors thank the Ferrol Industrial Campus and their excellence recruitment program "Research talent," launched in spring 2022, for the opportunity to do research and write this article.

KEYWORDS

- **hemostatic sponge**
- **chitosan**
- **konjak glucomannan**
- **graphene**
- **ionic gelation method**
- **porosity**
- **microparticles**

REFERENCES

1. Gu, R. L.; Sun, W. Z.; Zhou, H.; Wu, Z.; Meng, Z. Y.; Zhu, X. X.; Tang, Q.; Dong, J.; Dou, G. F. The Performance of a Fly-larva Shell-derived Chitosan Sponge as an Absorbable Surgical Hemostatic Agent. *Biomaterials* **2010,** *31,* 1270–1277.
2. Janvikul, W.; Uppanan, P.; Thavornyutikarn, B.; Krewraing, J.; Prateepasen, R. *In vitro* comparative Hemostatic Studies of Chitin, Chitosan, and their Derivatives. *J. Appl. Polym. Sci.* **2006,** *102,* 445–451.
3. Igai, H.; Yamamoto, Y.; Chang, S. S.; Yamamoto, M.; Tabata, Y.; Yokomise, H. Trachea Cartilage Regeneration by Slow Release of Basic Fibroblast Growth Factor from a Gelatin Sponge. *J. Thorac. Cardiovasc. Surg.* **2007,** *134,* 170–175.
4. Schonauer, C.; Tessitore, E.; Barbagallo, G.; Albanese, V.; Moraci, A. The Use of Local Agents: Bone Wax, Gelatin, Collagen, Oxidized Cellulose. *Eur. Spine J.* **2004,** *13,* S89–S96.
5. Björses, K.; Holst, J. Various Local Hemostatic Agents with Different Modes of Action; an *in vivo* Comparative Randomized Vascular Surgical Experimental Study. *Eur. J. Vasc. Endovasc. Surg.* **2007,** *33,* 363–370.
6. Boateng, J. S.; Matthews, K. H.; Stevens, H. N. E.; Eccleston, G. M. Wound Healing Dressings and Drug Delivery Systems: A Review. *J. Pharm. Sci.* **2008,** *97,* 2892–2923.
7. Song H., Chen, A.; Wang, S. h.; Kang Y.; Ye, S.; Liu Y.; Wu, W. Preparation of Chitosan-Based Hemostatic Sponges by Supercritical Fluid Technology. *Mater. (Basel).* **2014,** *7,* 2459–2473.
8. Anaya, P.; Cárdenas, G.; Lavayen, V.; García, A.; O'Dwyer, C. Chitosan Gel Film Bandages: Correlating Structure, Composition, and Antimicrobial Properties. *J. Appl. Polym. Sci.* **2013,** *128,* 3939–3948.
9. Zhang, J. L.; Xia, W. S.; Liu, P.; Cheng, Q. Y.; Tahi, T.; Gu, W. X.; Li, B. Chitosan Modification and Pharmaceutical/Biomedical Applications. *Mar. Drugs* **2010,** *8,* 1962–1987.

10. Jayakumar, R.; Prabaharan, M.; Sudheesh Kumar, P. T.; Nair, S. V.; Tamura, H. Biomaterials Based on Chitin and Chitosan in Wound Dressing Applications. *Biotechnol. Adv.* **2011,** *29,* 322–337.
11. Freitas, R.; Spin-Neto, R.; Spolidório, L. C.; Campana-Filho, S. P.; Marcantonio, R. A. C.; Marcantonio, E., Jr. Different Molecular Weight Chitosan-based Membranes for Tissue Regeneration. *Materials* **2011,** *4,* 380–389.
12. Millner, R. W. J.; Lockhart, A. S.; Bird, H.; Alexiou, C. A New Hemostatic Agent: Initial Life-saving Experience with Celox (chitosan) in Cardiothoracic Surgery. *Ann. Thorac. Surg.* **2009,** *87,* e13–e14.
13. Nwe, N.; Furuike, T.; Tamura, H. The Mechanical and Biological Properties of Chitosan Scaffolds for Tissue Regeneration Templates are Significantly Enhanced by Chitosan from Gongronella Butleri. *Materials* **2009,** *2,* 374–398.
14. Feng, C.; Khulbe, K. C.; Matsuura, T. Recent Progress in the Preparation, Characterization, and Applications of Nanofibers and Nanofiber Membranes via Electrospinning/ Interfacial Polymerization. *J. Appl. Polym. Sci.* **2010,** *115,* 756–776.
15. Jiao, Y.; Liu, Z.; Zhou, C. Fabrication and characterization of PLLA-chitosan Hybrid Scaffolds with Improved Cell Compatibility. *J. Biomed. Mater. Res. Part A* **2007,** *80,* 820–825.
16. Hoang, N. H.; Le Thanh, T.; Sangpueak, R.; Treekoon, J.; Saengchan, C.; Thepbandit, W.; Papathoti, N. K.; Kamkaew, A.; Buensanteai, N. Chitosan Nanoparticles-based Ionic Gelation Method: A Promising Candidate for Plant Disease Management, *Polymers* **2022**, *14,* 662.
17. Fan, W.; Yan, W.; Xu, Z.; Ni, H. Colloids and Surfaces B: Biointerfaces Formation Mechanism of Monodisperse, Low Molecular Weight Chitosan Nanoparticles by Ionic Gelation Technique, *Colloids Surf. B Biointerf.* **2012,** 9021–27. https://doi. org/ 10.1016/j.colsurfb.2011.09.042
18. Wonnie Ma I. A.; Ammar, S. H.; Bashir S.; Selvaraj, M.; Assiri, M. A.; Ramesh, K.; Ramesh, S. Preparation of Hybrid Chitosan/Silica Composites via Ionotropic Gelation and its Electrochemical Impedance Studies, *Prog. Org. Coat.* **2020,** *145,* 105679.
19. Jewelewicz, D. D.; Cohn, S. M.; Crookes, B. A.; Proctor, K. G. Modified Perihepatic Packing; A Creative and Beneficial Method for Management of High Grade Liver Injury *J. Trauma* **2003,** *55,* 275–280.
20. Amoozgar, H.; Abtahi, S.; Edraki, M.R. et al. Iran. J. Pediatr. Effect of Celox Powder on Initial Hemostasis After Cardiac Catheterization in Pediatric Patients with Congenital Heart Disease: A Prospective Study. *Iran. J. of Pediatr.* **2020,** *6,* 30.
21. Otrocka-Domagala, I.; Jastrzebski, P.; Adamiak, Z. et al., Safety of the Long-term Application of QuikClot Combat Gauze, ChitoGauze PRO and Celox Gauze in a Femoral Artery Injury Model in Swine—a Preliminary Study *Pol. J. Vet. Sci.* **2016,** *19,* 337–343.
22. Shakiba-Marani, R.; Ehtesabi, H. A Flexible and Hemostatic Chitosan, Polyvinyl Alcohol, Carbon Dot Nanocomposite Sponge for Wound Dressing Application. *Intern. J. of Biol. Macromol.* **2023,** *224,* 831–839.
23. Xia, G. et al., Preparation of a Novel Asymmetric Wettable Chitosan-based Sponge and its Role in Promoting Chronic Wound Healing, *Carbohydr. Polym.* **2020,** *227* (115), 296.

24. Synthesis and Characterization of ZnO Nanorods and Nanodisks from Zinc Chloride Aqueous Solution *Nanoscale Research Letters* Том **2009,** *4* (3), 247–253. https://doi.org/10.1007/s11671-008-9233-2.
25. Kang, P. L.; Chang, S. J.; Manousakas, I.; Lee, C. W.; Yao, C. H.; Lin, F. H.; Kuo, S. M. Development and Assessment of Hemostasis Chitosan Dressings. *Carbohydr. Polym.* **2011,** *85,* 565–570.
26. Rivero, S.; García, M. A.; Pinotti, A. Crosslinking Capacity of Tannic Acid in Plasticized Chitosan Films *Carbohyd. Polym.* **2010,** *82* (2), 5, 270–276.

CHAPTER 11

Obtaining and Studying the Properties of Nanocomposites Based on a Mixture of Polyethylenes of High and Low Pressure with Metal-Containing Nano Fillers

N. I. KURBANOVA[1], G. H. MAMEDOVA[2], and T. M. GULIYEVA[1]

[1]*Institute of Polymer Materials of the Ministry of Science and Education of the Republic of Azerbaijan, Sumgait, Azerbaijan*

[2]*Azerbaijan State Oil and Industry University, Azerbaijan*

ABSTRACT

The properties and structure of the nanocomposite obtained by the effect of nanofillers containing copper(I) oxide nanoparticles stabilized in a high-pressure polyethylene matrix and a maleinated high-pressure PE matrix on a mixture of high- and low-pressure polyethylene were studied by the DTA method. The result shows that a small amount of nanofiller included in the polymer is located at the boundary of the interphase layer of the structural elements of HPPE and LPPE and helps to form heterogeneous nucleation centers, which leads to the formation of a relatively fine spherulitic structure. It leads to the improvement of the physical-mechanical and thermal properties of the resulting nanocomposite.

Advances in Engineering Materials: Preparation for Sustainable Process Development. Tamara Tatrishvili and A. K. Haghi (Eds.)

11.1 INTRODUCTION

It is known that the modification method consisting of the creation of polymer–polymer compositions is widely used to improve directly the properties of industrial polymers.[1] In modern polymer materials science, polymer mixtures are used to produce composite materials with a complex of improved functional properties, as well as to expand their range and areas of application. Due to the mixing of polymers, many properties of the resulting composites can be improved, primarily mechanical strength. The complex technological and operational properties of the obtained materials are determined by the structure of polymer mixtures. At the same time, the uniqueness of polymer mixtures lies in the fact that they not only retain the properties of the original components of the mixture but can also acquire new properties that none of the original components possess. However, even in the absence of synergy, that is, exceeding the obtained properties compared to their additive values, a mixture of polymers in one composite combines the basic properties of both components.[2] Metal-containing nanofillers create wide opportunities for determining the properties of materials without significantly changing their composition.[3] The use of nanoparticles of metals with variable valency (copper, cobalt, etc.) in polymers allows for obtaining fundamentally new materials that are widely used in radio and optoelectronics, such as magnetic, electrical conductors, and optical media.[4–10]

11.2 EXPERIMENTAL METHODS AND MATERIALS

11.2.1 HPPE/LPPE

HPPE brand 15803-020, $\rho = 0.917–0.921$, MFR 1.5–2.5 g · 10-1 min ($T = 190$ °C, load 2.16 kg); LPPE brand HM0349PE, $\rho = 0.949$, MFR 8.3 g 10-1 min ($T = 190$°C, load 21.6 kg).

11.2.2 HPPE/LPPE/HPPE NF

Nanoparticles containing copper oxide stabilized by a polymer matrix were used as nanofillers. The nanofiller (NF) contains copper-containing nanoparticles (NPs) stabilized by a matrix of high-pressure polyethylene

(HPPE) obtained by a mechano-chemical method. The composition of used nanoparticles is 5.0 mass %, the size is 35 ± 1.0 nm, and the degree of crystallinity is 25–45%.[11,12] The ratio of primary components (mass %): HPPE/LPPE/HPPE NF = 50/50/ (0; 0.3; 0.5; 1.0). The initial components were mixed on laboratory rollers at 150 °C for 15 min. In order to conduct mechanical tests, the resulting mixtures were pressed at 190 °C to a thickness of 1 mm and at a pressure of 10 MPa for 10 min.

11.2.3 HPPE/LPPE/MPE NF

The copper-containing nanoparticles stabilized by the polymer matrix were used as nanofillers. The NF contains the copper-containing nanoparticles stabilized by a matrix of maleinated HPPE of firm "OLENTENA," obtained by the mechanical-chemical method in a polymer melt. The composition of used nanoparticles is 5.0 mass%, the size is 25 ±1.0 nm, and the degree of crystallinity is 25–45%.[13] The ratio of the components of the composition (mass %): HPPE/LPPE/MPE NF = 50/50/ (0; 0.3; 0.5; 1.0). The nanocomposite polymer materials were obtained by mixing HPPE and LPPE with the copper-containing nanofiller on laboratory rollers at a temperature of 150°C for 15 min. For carrying out the mechanical testing, the obtained mixtures were pressed as plates with a thickness of 1 mm at 190°C and pressure of 10 MPa for 10 min.

11.3 RESULTS AND DISCUSSION

11.3.1 PHYSICAL AND MECHANICAL PROPERTIES OF HPPE/LPPE/HPPE NF

Nanocomposite polymeric materials based on HPPE and LPPE with copper-containing nanofillers have been obtained. The physical-mechanical, thermophysical, and thermal properties of the obtained nanocomposites have been studied. Table 11.1 shows the physical and mechanical parameters of the obtained composite materials.

Table 11.1 shows that the addition of 0.5 wt. % nanofiller to the composition increases its strength limit from 16.69 to 19.77 MPa. Adding more than 0.5% of the nanofiller to the composite is followed by a decrease in the strength limit, which is most likely related to the aggregation of nanoparticles and causes the formation of micro defects in the structure of the polymer. The addition of 0.5% of nanofiller increases the relative elongation at break by 1.7 times, which is explained by the simultaneous increase of mechanical strength and relative elongation based on the synergistic effect of the HPPE/LPPE mixture of copper oxide particles in the MPE matrix. According to Vike, the heat resistance increases from 155 to 175°C.

TABLE 11.1 Physical and Mechanical Parameters of the Obtained Nanocomposites.

Composites		Tensile strength at break, MPa	Elongation, %, ε_S	Heat resistance according to Vikat, °C
HPPE/LPPE	(50/50)	16.69	590	155
HPPE/LPPE/ $NHCu_2O$/HPPE	(50/50/0.3)	17.69	959	165
	(50/50/0.5)	19.77	1000	175
	(50/50/1.0)	17.78	690	160
HPPE/LPPE/ $NHCu_2O$/MPE	(50/50/0.3)	17.07	920	160
	(50/50/0.5)	19.15	996	175
	(50/50/1.0)	19.01	951	162

11.3.2 THERMAL INDICATORS OF THE STUDIED SAMPLES OF LDPE/HDPE AND LDPE/HDPE NF

Since nanocomposite samples with 0.5 mass% nanofiller containing nanoparticles of copper(I) oxide showed the best physical and mechanical properties, the thermal stability of this sample was studied. The thermal stability of the studied samples of mixtures based on HPPE/LPPE containing NF with copper(I) oxide NPs was estimated from the activation energy of thermooxidative destruction, calculated by the double logarithm method from the TGA curve.[13]

The result of DTA is shown in Table 11.2.

As can be seen from the data in Table 11.2, the addition of NF-containing copper oxide nanoparticles into the composition of mixed polyethylenes contributes to a significant increase in the decomposition temperature of the samples: up to 425°C, no mass loss is observed for both samples under study; at 450°C, a mass loss of 1 is observed %; at 470°C—2.5%; and at 480°C—4.0%; and at 490°C—8.0% for the initial mixture; while for a composite containing NF with copper(I) oxide NPs, at 450°C, a

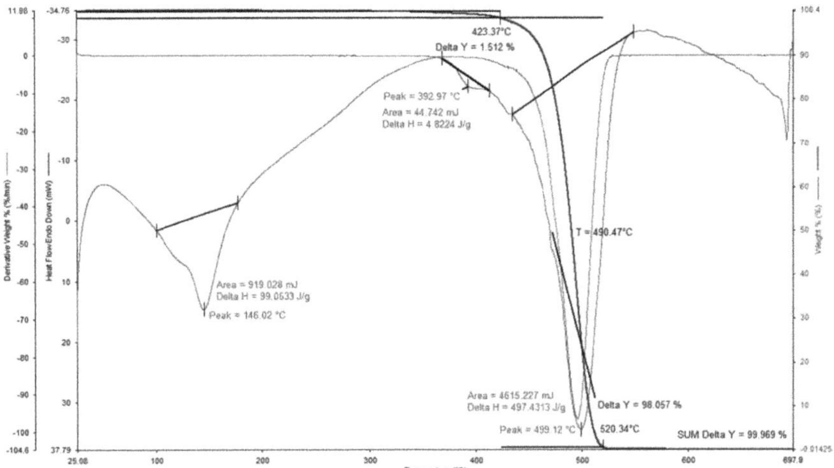

FIGURE 11.1 Thermogram of a sample of HPPE/LPPE.

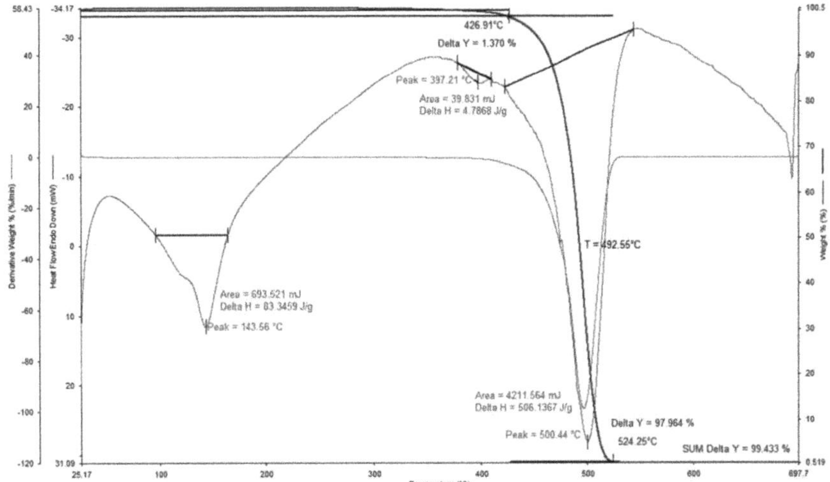

FIGURE 11.2 Thermogram of a sample of HPPE/LPPE/HPPE NF.

mass loss of 1% is observed; at 470°C—2.0%; at 480°C—3.0%; and at 490°C—6.0%. The activation energy (E_a) increases from 232 to 255 kJ/mol. The introduction of NF-containing copper(I) oxide NPs into the composition of mixed polyethylenes contributes to a significant increase in the decomposition temperature of the samples: up to 400°C, no weight loss is observed for both samples under study; at 400°C, a weight loss of

1% is observed; at 450°C—1.0%; at 470°C—2.5%; at 480 °C—2.0%; and at 490°C—8.0% for the initial mixture; while for the composite containing NF with copper(I) oxide NPs, at this temperature, the weight loss is only 4.0%, which is two times less. The activation energy (E_a) of the decomposition of the thermal oxidative destruction of the obtained nanocomposite increases from 232.96 to 274.76 kJ/mol.

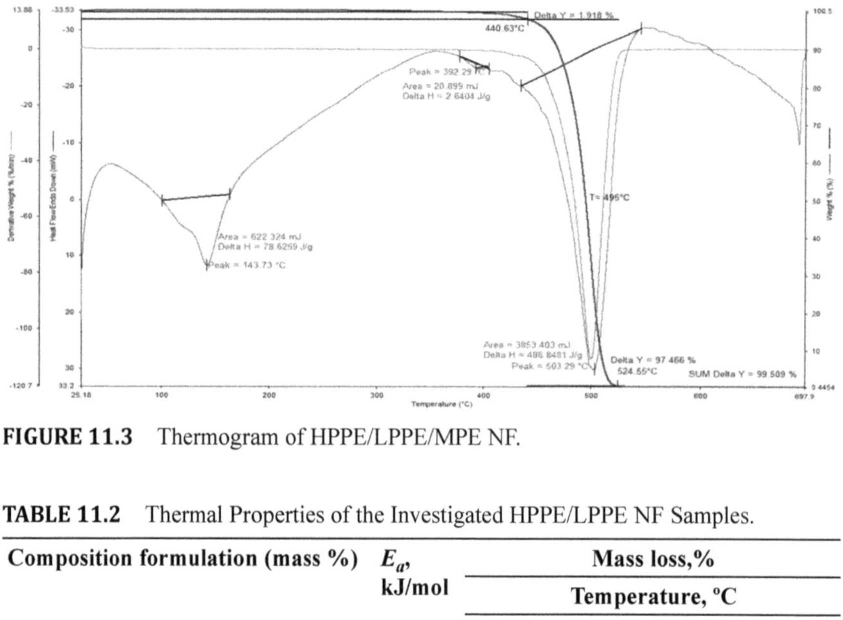

FIGURE 11.3 Thermogram of HPPE/LPPE/MPE NF.

TABLE 11.2 Thermal Properties of the Investigated HPPE/LPPE NF Samples.

Composition formulation (mass %)		E_a, kJ/mol	Mass loss,%				
			Temperature, °C				
			425	**450**	**470**	**480**	**490**
HPPE/LPPE	(50/50/0)	232	0	1.0	2.5	4.0	8.0
HPPE/LPPE/NPsCu$_2$O/HPPE	(50/50/0.5)	255	0	1.0	2.0	3.0	6.0
HPPE/LPPE/NPsCu$_2$O/MPE	(50/50/0.5)	274	0	0.5	1.5	2.0	4.0

As a result of thermal analysis, the thermal stability of nanocomposites obtained by adding nanofiller containing copper(I) oxide particles to the polyolefin mixture was improved.

Mechanical strength, relaxation, and other properties of polymer–polymer and polymer–filler mixtures are mainly explained by the concept of an interphase layer.[14, 15]

The metal-containing nanoparticles used in the work are located at the boundary of the interphase layer of the structural elements of HPPE and LPPE, helping to form heterogeneous nucleation centers, which leads to the formation of a relatively fine spherulitic structure. It leads to the improvement of physical-mechanical and thermal properties of the resulting nanocomposite.

The bactericidal properties of the copper(I) nanoparticles used in the article, stabilized by the HPPE and MPE matrix, have been revealed.

The high antibacterial activity of copper-containing nanocomposites toward Gram-positive and Gram-negative bacteria was revealed and also the absence of the probability of the release of copper(I) oxide nanoparticles into the environment in accordance with environmental materials safety requirements was proved.[16] Therefore, it shows the possibility of using them as antimicrobial additives to remove harmful microbes from drinking water, as well as additives to polymeric materials to give them bactericidal properties.

The obtained nanocomposites with improved properties can be recommended for use as a promising structural material in mechanical engineering, electrical engineering, petrochemistry, and construction; as an antibacterial in the medicine and food industry; and for radiation protection against neutron radiation.

11.4 CONCLUSION

The effect of nanofiller containing copper(I) oxide nanoparticles, stabilized by a matrix of HPPE and MPE, on the properties of the HPPE/LPPE-based composites was studied. Adding more than 0.5% of the nanofiller to the composite is followed by a decrease in the strength limit, which is most likely related to the aggregation of nanoparticles and causes the formation of micro defects in the structure of the polymer. The addition of 0.5% of nanofiller increases the relative elongation at break by 1.7 times, which is explained by the simultaneous increase of mechanical strength and relative elongation based on the synergistic effect of the HPPE/LPPE mixture of copper oxide particles in the MPE matrix.

The thermal properties of the studied samples showed that the activation energy of the decomposition of the thermal oxidative degradation of the obtained nanocomposite increases from 232 to 255 kJ/mol (HPPE NF) and from 232 to 274 kJ/mol (MPE NF). The metal-containing nanoparticles used in the work are located at the boundary of the interphase layer of the structural elements of HPPE and LPPE, helping to form heterogeneous nucleation centers, which leads to the formation of a relatively fine spherulitic structure. It leads to the improvement of physical-mechanical and thermal properties of the resulting nanocomposite.

The bactericidal properties of the copper(I) nanoparticles used in the article, stabilized by the HPPE and MPE matrix, have been revealed. The obtained nanocomposites with improved properties can be recommended for use as a promising structural material in mechanical engineering, electrical engineering, petrochemistry, and construction; as an antibacterial in medicine and food industry; and for radiation protection against neutron radiation.

KEYWORDS

- **high-pressure polyethylene**
- **low-pressure polyethylene**
- **nanoparticles of copper(I) oxide**
- **strength**
- **deformation**
- **thermal properties**
- **DTA analyzes**

REFERENCES

1. Ermakov, S. N.; Kravchenko, T. P. Polymer Compatibility. Thermodynamic and Chemical Aspects. *Plastics*. **2012,** *4,* 32–38.
2. Kuleznev, V. N. *Blends and Alloys of Polymers. Lecture Notes*; St. P.: Scientific Foundations and Technologies, 2013, p 216.
3. Nikolaychik, Y. U. A.; Kuis, D. V.; Svidunovich, N. A.; Rovin, S. L. General Problems of Development and Implementation of Nanomaterials and Nanotechnologies. *Casting and Metallurgy*, **2020,** *4,* 152–162.
4. Kurbanova, N. I.; Ragimova, S. K.; Alimirzoeva, N. A.; Ishenko, N. Y. A.; Medyakov, V. V. Composites Based on Isotactic Polypropylene and High-pressure Polyethylene with Zinc-Containing Nanofillers. *Inorgan. Mater.: Appl. Res.*, **2022,** *13* (2), 485–488.

5. Kurbanova, N. I.; Guliyeva, T. M.; Ishenko, N. Y. A. Preparation and Properties of Metal-Containing Nanocomposites Based on Isotactic Polypropylene and Butadiene–Acrylonitrile Rubber. *Russian J. Appl. Chem.* **2021,** *94* (1), 17–21.
6. Allahverdiyeva, K. H. V.; Kakhramanov, N. T.; Arzumanova, N. B.; Abdalova, S. R. Thermal Strain Properties of Metal-Polymer Composites Based on Low-Density Polyethylene and Copper. *Inorganic Mater.: Appl. Res.* **2022,** *13* (1), 116–120.
7. Alimirzayeva, N. A. The Effect of Metal-containing Nanoparticles on the Thermo-oxidative Properties of Thermoplastic Elastomers Based on Isotactic Polypropylene and Ethylene Propylene Diene Rubber. *New Mater. Comp. Appl.* **2022,** *6* (3), 220–229.
8. Kurbanova, N. I., Guliyeva, T. M., Ishenko, N. Y. A. Properties of Nanocomposites Based on Isotactic Polypropylene and High-Pressure Polyethylene with Metal-Containing Nanofillers. *Inorganic Mater.: Appl. Res.* **2021,** *12* (2), 457–460.
9. Guliyeva, T. M. Zinc-containing Nanocomposites on the Basis of High Pressure Polyethylene. *Az. Chem. J.* **2020,** *2,* 34–38.
10. Kurbanova, N. I., Guliyeva, T. M., İshenko, N. Y. A. Properties of Nanocomposites of High-Pressure Polyethylene with Metal-Containing Nanofillers. *Inorganic Mater.: Appl. Res.* **2020,** *11* (5), 1180–1183.
11. Kurbanova, N. I.; Alimirzoeva, N. A.; Guseinova, Z. N.; Nurullayeva, D. R. Ecological Method of Preparation of Metal-Containing Nanoparticles in Polyethylene Matrix. *Book of Proceedings of ITWCCST*, **2017,** 24.
12. Kurbanova, N. I.; Aliyev, A. T.; Guliyeva, T. M.; Ragimova, C. K.; Axmadbekova, C. F.; Ishenko, N. Y.; Nurullayeva, D. R. Metal-Containing Nanoparticles in Maleinized Polyethylene Matrix. *Mater. PolyChar. 26 World Forum on Adv. Mater.* **2018,** 59.
13. Kryzhanovsky, V. K. Technical Properties of Polymer Materials: Educational and Reference Manual/Under the General Editorship of Prof. V.K. Kryzhanovsky. *St. Petersburg: Profession,* **2007.** 240.
14. Pomogailo, A. D., Molecular Polymer–polymer Compositions. Synthetic Aspects, *Russ. Chem. Rev.* **2002,** *71* (1), 5–38.
15. Kahramanli, Y. U. N. Incompatible Polymer Mixtures and Composite Materials Based on them. *Baku: Elm.* **2013.** 152.

CHAPTER 12

Morphological Investigation and Fractal Properties of New Inorganic-Based Nano Materials and Nanocarriers

SHAHRIAR GHAMMAMY, FARZANE YOUSEFI RAMANDI, MELINA BABOLHAVAEJI, NILOOFAR PALIZDAR, and ALI MIRHASHEMI

Department of Chemistry, Faculty of Science, Imam Khomeini International University, Qazvin, Iran

ABSTRACT

Many events, especially the surrounding and natural events, can be explained and interpreted by combining the nano theory and a special kind of geometry known as fractal geometry. In the field of nanotechnology, it is attainable to create new and more effective drugs that have enhanced physical and chemical characteristics, such as high reactivity and the capacity to select the cell membrane. Many fields are exciting at the beginning of this category for studying to enhance the fundamental quality of people's lives. Nanoparticles have the potential to overcome many drug design deficiencies and delivery problems. Metallic and nonorganic nanoparticles have various usages in drug releasing mechanisms and are much sought after for their multifarious aspects such as abroad approachability, exact bio fitting, and high level of out-layer activity. Many efforts have undertaken to expand the drug delivery systems. Usage of nanoparticles in drug delivery systems shows promising scope for effective and exact drug releasing. Layered double hydroxide (LDH) is a newly proposed material for drug delivery. For its

Advances in Engineering Materials: Preparation for Sustainable Process Development. Tamara Tatrishvili and A. K. Haghi (Eds.)

experimental and practical usages, it is important that it shows the ability to control the amount of the drug and release rate in living unites (in vivo). In this research, some of the new nano compounds were synthesized. These new materials were identified by physical and spectral techniques such as FT-IR and UV spectra, X-ray diffraction, SEM techniques, and some physical properties. The biological properties such as anticancer activities of these nano compounds against a different kinds of tumor cells such as panel of human tumor cell lines have been studied. Fractal geometry has recently been recognized by us and other researchers as a suitable analytical tool for describing nanostructures and nanomaterials. As known, fractals are infinitely complex patterns that are similar at different scales. Their morphological properties have been studied and a detailed calculation and study of morphological properties of a typical nanocarrier have been shown in this manuscript.

12.1 INTRODUCTION

Nanomaterials are defined as materials in the nanodimension range of 1–100 nm which possess unique properties such as optical, magnetic, and electrical properties. Nanomaterials used in cancer therapy and treatment can be classified into several main categories. Double hydroxide layered materials are special compounds that contain layers, especially oxide compounds. These nanomaterials can be used in pharmaceutical applications as nanocarriers. The new kind of nanocarriers contains layered double hydroxides (LDHs), which are composed of block wall-like layers, as famous as anion clays, and are widely used in catalysis and anion exchange processes. A wide variety of differently charged anions can be incorporated into the double-layer structure, these anions and cations may contain halides (chlorides) and anions (carbonates, nitrates). Many texts have been written about the application of nanoparticles (NPs) and nanomaterials for drug delivery to target tissues and cells with selective releasing amounts.[1–4]

Nanotechnology has advanced on all fronts: instruments, goods, literature, and profits. According to the project on engineering nanotechnologies, it is noted that nanomaterials exhibited special and extreme temperature and conductivity, and surface properties. On the basis of these unique features, the application of nanomaterials is very extensive. In this research, we investigated the fractal properties of nanomaterials. By using nanotechnologies, it is possible to find and use more effective methods in drug delivery systems so that better results can be achieved with less drug consumption. As is known in medication, especially on a nanoscale, the self-similarity behavior of particles can be seen and found. Self-similarity is one of the most important features of fractal science. One of the branches of mathematics is fractal geometry which has shown a powerful ability to explain and simulate different forms of nature. Fractal geometry has the ability to interpret chaotic and apparently disordered sets. Based on fractal dimension analysis and by using the image processing method, abundant and powerful information can be obtained in the morphological field of nanomedicines. Fractals are irregular and form suddenly with equal dissimilarities in all amounts. The fractal dimension is an important integer for the study of curvy forms and multiplicity of pictures and subjects in contrast to the widespread numbered Euclidian dimension that can be a real number. By calculating and finding the fractal dimension, the future behavior of data and fractal geometry can be estimated better.[5–8]

In this paper, the fractal and morphological parameters of typical nanocarriers have been studied and calculated in detail.

12.2 EXPERIMENTAL METHODS AND MATERIALS

Nanomaterials were characterized by the FT-IR model Bruker-Tensor 27 spectrophotometer. The product surface morphology was studied by a SEM. For more characterization and study, XRD measurements were performed. For the synthesis of nanoparticles, all chemicals were obtained from Merck. Figure 12.1 shows the infrared spectrum of a typical nanocarriers that displays several vigorous and sharp transmittance bands that arise from several functional groups existing in the molecules. The comparison of nano materials with raw materials shows that drug loading has taken place among the nanocarrier layers. The experimental PXRD pattern of nanoparticle (Figure 12.2) was in good agreement with nano size of matter and exhibited high crystallinity. This pattern also agreed with anions in adopting a nearly perpendicular location according to the nanocarriers. The identification of anionic, such as size, load, and intercalation space of the gallery interlayers, can be specified by this method.[9–11]

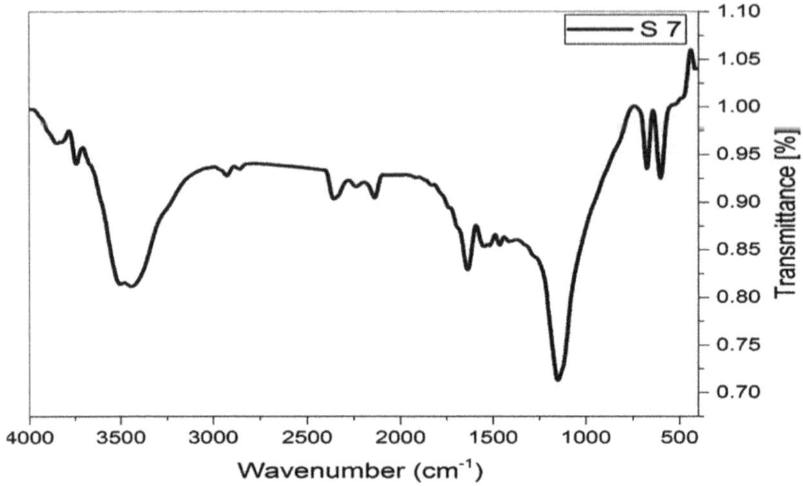

FIGURE 12.1 IR spectrum of a typical nanocarrier.

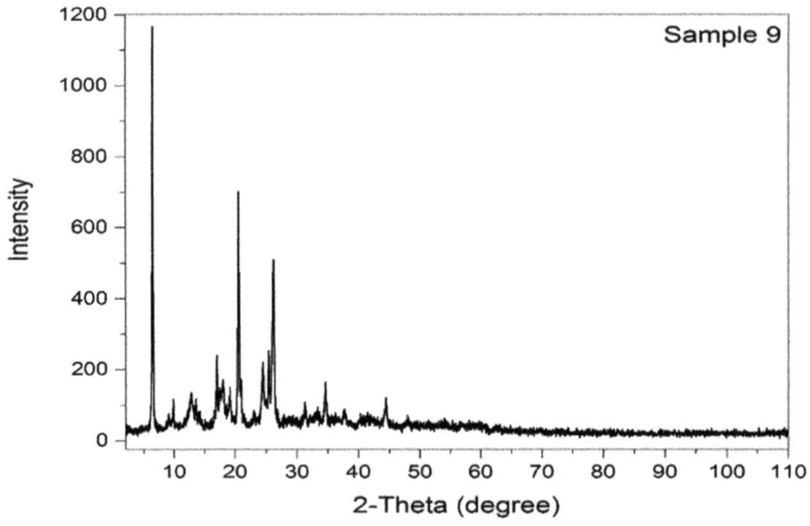

FIGURE 12.2 XRD powder diffraction of nanocarrier.

12.2.1 COMPUTATIONAL METHOD

By using MATLAB software, ten (10) section numbers of the FESEM pictures of nano inorganic compounds were selected and studied. Their fractal dimensions were calculated. As an example, the parts were selected from the FESEM image of an oxide after attacking the bacterial tissue, and its fractal dimensions were calculated and found (Figure 12.3). The obtained fractal dimensions' data from MATLAB software were fed into the SPSS software as initial data for calculating distribution normality, data relations, standard deviation, mean of integers, data frequency behavior, and variance. According to the obtained numbers, we drew a logarithmic graph with the vertical axis of log ($1/r$) and horizontal (N/r) log.

We obtained the fractal dimension by using programming in MATLAB. There are different methods to obtain the fractal dimension. We used the box-counting method. The method of calculating the fractal dimension with this method is as follows:

Consider the following square, which can be broken into N-like pieces.

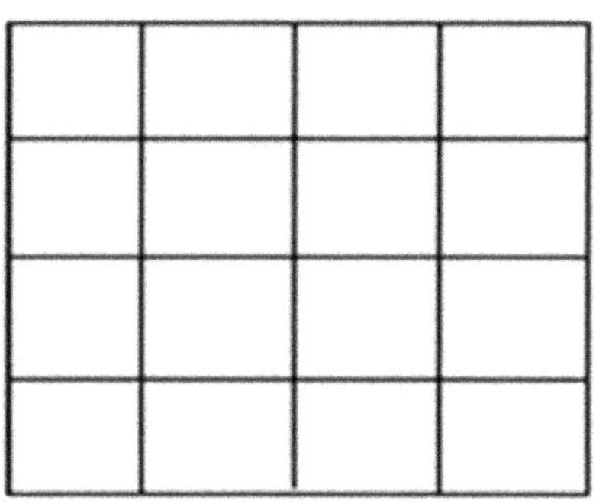

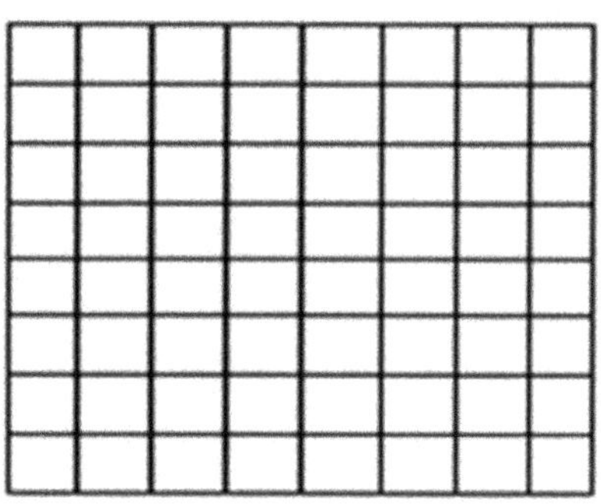

Each slice is expressed by the magnification factor N. Therefore:

$$D = \lim_{r \to 0} \frac{\log Nr}{\log(1/r)} \tag{12.1}$$

where D is the dimension of fractal and N_r is fractal dimensions' squares including part number.

$$\mathit{Dimension} = \frac{\log(\text{number of self-similar pieees})}{\log(\text{magnitication factor})} \tag{12.2}$$

The fractal dimension is obtained from the following equation:

$$\mathit{Fractal\ Dimension} = \frac{\log(\text{number of self-similar pieees})}{\log(\text{magnitication factor})} \tag{12.3}$$

Now we can get the dimensions of different fractals, for example, for the famous Serpinski triangle, which consists of three self-similar pieces, each with a magnification factor of 2. The dimension is:

$$\frac{\log(\text{number of self-similar pieees})}{\log(\text{magnitication factor})} = \frac{\log 3}{\log 2} = 1.53 \quad (12.4)$$

By using the SPSS statistical analysis software, we obtained the correlation coefficients and regression coefficients of the points of the diagram of fractal dimension and the mathematical equation related to each image.

12.2.2 TYPICAL ANALYSIS OF THE FRACTAL DIMENSIONS OF A NANO INORGANIC COMPOUNDS

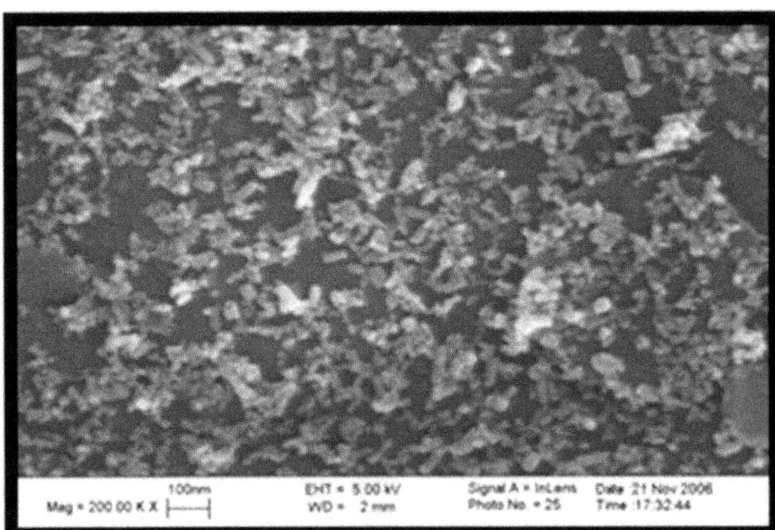

FIGURE 12.3 SEM image of a nano inorganic compound (1).

According to this diagram, the fractal dimension of this nanomaterial is FD = 1.8537.

In order to obtain the correlation coefficient between the points on the line in the graph, we entered the data from Table 12.1 into the SPSS software and used the Pearson correlation coefficient. We obtained the correlation coefficient of these points. The correlation coefficient is shown in Table 12.2. Its amount is −0.999, which is a sign of the indirect relationship between Log(res) and Log(boxnum) (Figure 12.4).

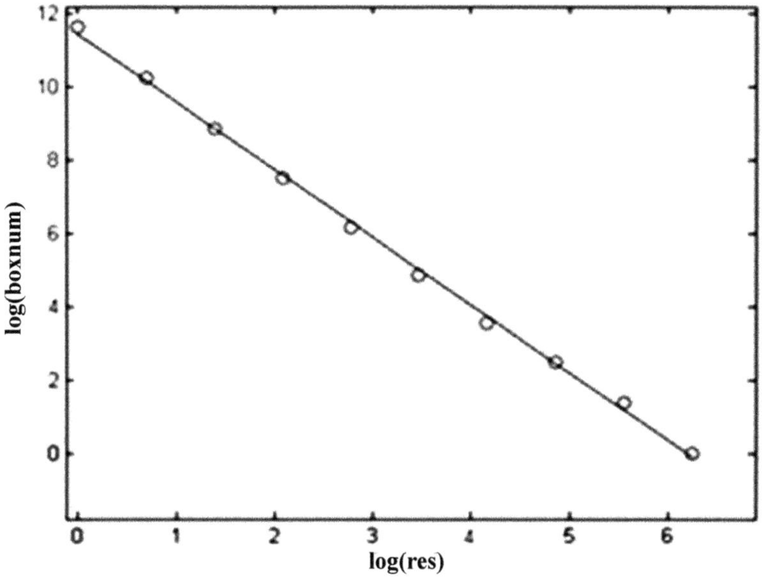

FIGURE 12.4 Sample graph of the results of the fractal dimension analysis of a nano inorganic compound counter box (1).

TABLE 12.1 The Spatial Value of the Points on the Fractal Dimension Diagram of Nano Inorganic Compound.

LOGres	LOGboxnum
0	11.64
0.6931	10.26
1.386	8.875
2.079	7.504
2.773	6.165
3.466	4.836
4.159	3.555
4.852	2.485
5.545	1.386
6.238	0

The correlation coefficient is a statistical tool to determine the type and degree of relationship of one quantitative variable with another quantitative variable. The correlation coefficient is one of the criteria used to determine the correlation between two variables, which can show the correlation and relationship between data and also the type of relationship (direct or inverse). This coefficient amount is between 1 and −1 and is equal to zero if there has not any relationship between two variables. In this article, Pearson's correlation coefficient is used, which can be used for the calculation of quantitative variables, where it is assumed that continuous variables tarenormally distributed, this equation is strongly related and affected by sample size. So, we checked more points to make sure we applied a powerful test. Pearson's correlation coefficient is based on covariance and deviation variables, where their estimation can be used to calculate Pearson's correlation coefficient. As Table 12.2 shows, the fractal dimension correlation of nano inorganic compounds is 0.999. The found coefficient indicates a strong relation between the various parts of the image, all of which confirm the fractal nature of nano inorganic compounds. In short, incredibly complex order can be seen in the seemingly disordered. After calculating the correlation coefficient of the points, we analyzed the data of Table 12.1 by using the regression coefficient of linear equations. On the basis of Table 12.3, the slope of the line was calculated as −1.854, which has a very small difference with the dimension calculated by MATLAB software. The origin of the graph was found to be 11.452. In Table 12.4, the maximum and minimum numbers and the uncertainty of the data was obtained.

TABLE 12.2 Regression Coefficient of a Typical Nano Inorganic Compound.

Model		Unstandardized coefficients		Standardized coefficients	t	Sig.	Collinearity statistics	
		B	Std. error	Beta			Tolerance	VIF
1	(Constant)	11.452	0.094		121.884	0.000		
	LOGres	−1.854	0.025	−0.999	−73.001	0.000	1.000	1.000

[a]*Dependent variable*: LOGboxnum.

Log (boxnum) = −1.854 Log(res) + 11.452.

In the sections obtained from nano inorganic compound, the extremums such as minimum and maximum of fractal dimensions were found as −0.11 and 11.45, respectively. On the base of these calculations, the average fractal dimension was 5.67.

On the base of statistics and probability theory, skewness reflects the degree of probability distribution symmetry. The presence or absence of distribution function symmetry is shown by Skewness. On the base of skewness data for a perfectly symmetrical distribution, the skewness amount is zero. For an asymmetric distribution with stretching toward higher quantities, the skewness integer sign is a positive integer; and for an opposite situation, asymmetric distribution with stretching toward smaller quantities, the skewness amount is negative integer.

Predictors: (Constant), LOGres, b. Dependent Variable: LOGboxnum

TABLE 12.3 Variance Analysis of Linear Regression of Nano Inorganic Compound.

Model	Sum of squares	df	Mean square	F	Sig.
Regression	136.200	1	136.200	5.329E3	0.000
Residual	0.204	89	0.026		
Total	136.404				

In the sections obtained from nano inorganic compounds, the two extremum amounts of dimensions of fractals were 0.11 and 11.45, respectively. The average fractal dimension was 5.67.

Skewness reflects the amount and degree of asymmetry of the probability distribution in statistics and probability theory. Skewness amount is a measure and sign of the presence of distribution function symmetry. For example, a perfectly symmetrical distribution has a skewness equal to zero; positive skewness belongs to asymmetric distribution with stretching toward higher quantities, and negative skewness related to an asymmetric distribution with stretching toward smaller quantities. In Table 12.4, you can observe the skewness to be 0.

TABLE 12.4 The Numbers of Maximum, Average, Minimum, and Uncertainty Data Related to Nano Inorganic Compound.

	Minimum	Maximum	Mean	Std. deviation	Skewness
Predicted value	−0.11089	11.45249	5.67060	3.890154	0
Residual	−0.19157	0.212272	0.00000	0.150725	0.0106
Std. predicted d value	1.486	1.486	0.000	1.000	0
Std. residual	1.198	1.328	0.000	0.943	0

Dependent variable: LOGboxnum

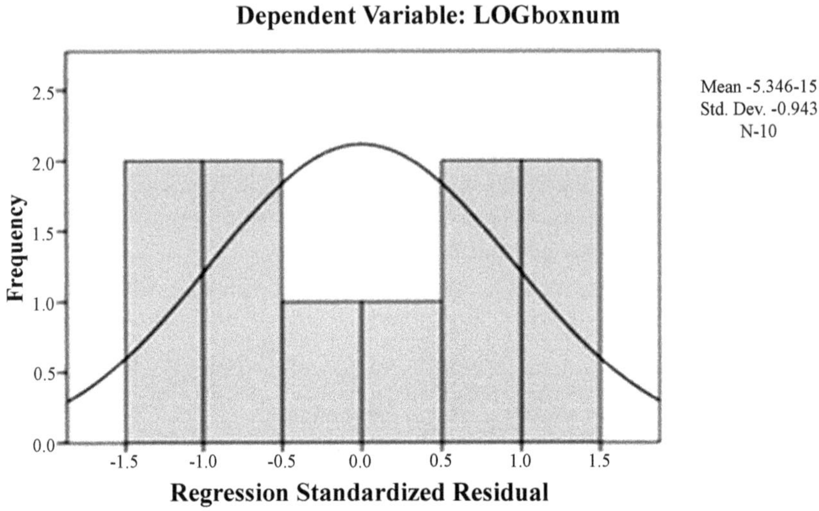

FIGURE 12.5 Histogram of a typical nano inorganic compound.

It shows a symmetrical distribution. Since the obtained figure is zero, there is symmetry. A gray-scale digital image was collected and formed from individual points of gray or luminance, before continuously varying tones. A typical image is separated and distributed into a number of specific lighting points, each of which parts are defined by a digital data value. In part, a pixel is the greatest amount of the picture. Each point of illumination is a pixel of the digital image. The histogram of an image shows the intensity distribution of pixels in the image and the ones that are consistently used as a reference. Interactive correction can be very effective and provide quick and accurate information. The histogram chart extracted with SPSS software is shown in Figure 12.5.

In a statistical population, the normality of the data can be obtained by using P–P plot diagrams. It seen that the data is scattered around a line, which is a sign of the normality of this statistical population. Normal and trendless P–P is shown in Figure 12.6. It was seen that the distribution of fractal dimensions of inorganic nanoparticles is in a straight and direct line and the hypothesis of normal distribution of errors has correct amount and behavior. Plotting normal P–P diagrams considering residuals is installed here. The points should be uniformly distributed if the fitted model is coherent and worked. The shape should be symmetrical and the points should be uniformly distributed. These represents a constant variance of errors that is optimized in natural phenomena. Summarizes of these process and calculations can help us to determination and extraction the fractal properties.

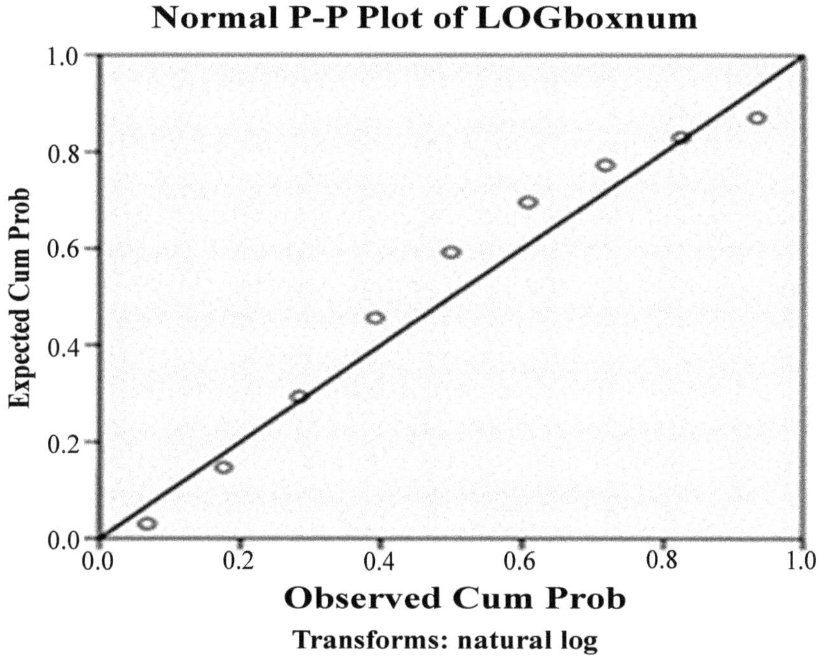

FIGURE 12.6 Normalized P–P diagram of a typical nano inorganic compound data.

12.3 RESULTS AND DISCUSSION

In this study, the fractal dimensions of nano inorganic compounds and their impact on tissues and cells were calculated and studied. For statistical analysis of the resulting data, the SPSS software was used. The primary aim of this research is to study and prove fractal behavior models in nano inorganic SEM pictures. Recently, scientists have come to the conclusion that some properties of nano inorganic compound, such as their conductivity and stability in the body, can be explained and interpreted using the characteristic of the fractal dimension. In the following, using MATLAB software, the degree of homogeneity or heterogeneity of nano inorganic compounds was checked using SEM images. This method is effective and can be used in some drug production companies for product grading. Eventually, some of the existing assumptions in the field of measuring the fractal dimension were discussed. According to the examination of the morphology of the SEM images in different stages of the work, it was confirmed that the antibacterial activity on the inorganic compound increases with the increase in the particle concentration or the decrease in the particle size.

Many fractal objects such as leaves, snowflakes, and trees are unique and special in nature. We know that the production of fractal nanoparticles with uniform and unique morphology and definite structures is difficult. In this part, to check the morphology of synthesized nanoparticles, the images recorded by electron microscope scanning (SEM) was investigated. Fractal dimensions obtained and extracted from MATLAB and SPSS softwares. With these softwares, the distribution normality, data relations, standard deviation, mean of integers, data frequency behavior, and variance analysis were analyzed. The study of the morphological properties explained with pictures has been done in this manuscript. The study of the morphological structure of the samples shows that the materials formed as nanopowders. Larger particles that are observed in the surface of the pictures were formed as a result of the aggregation and conjunction of the smaller particles, which manifested as irregular spherical bulky shapes. SEM images of nanoparticles are shown in Figure 12.5.[12,13] Morphological studies show similarity in nanoscale, as well as in bulk. Nanocarriers such as LDH compounds shown the layer systems and structure in SEM pictures. Their XRD patterns show layer structure by near 10 angles pattern. In the following picture, can see the layer pattern in SEM of a nanocarrier (Figure 12.7)

12.4 CONCLUSION

Recently, there has been an exponential increase in studies of using nanotechnology for tissue engineering applications. This paper only covered the recent progress using nanomaterials for medicine as a nanocarriers. The landscape of nanomaterials for medicines of the next generation is rich with options, and innovative solutions will likely be found in the wise combination of different components. It is clear that the examples reported in this viewpoint are only a few representative of a class of materials that is continuously expanding and that includes an exceedingly high number of examples and ideas. The fractal properties and morphological studies of nano compounds such as nanocarriers can be developed to extract more data from their spectroscopic data such as SEM pictures.

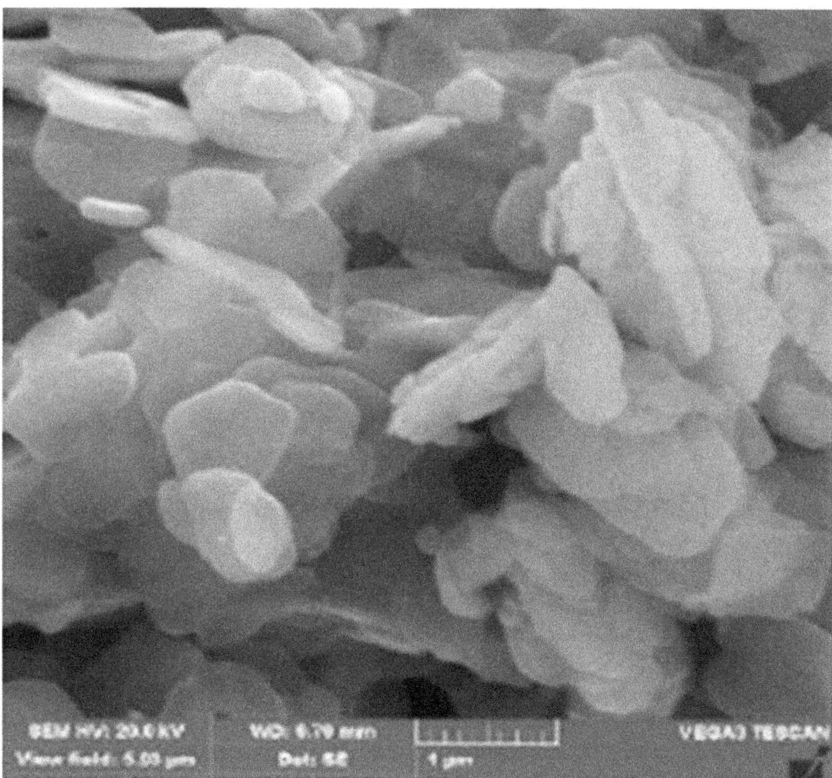

FIGURE 12.7 SEM of a typical LDH nanocarrier.

ACKNOWLEDGMENTS

The authors gratefully acknowledge support from the research council of Imam Khomeini International University.

KEYWORDS

- **morphology**
- **fractal properties**
- **nanomaterial**
- **nanocarriers**

REFERENCES

1. Ghamami, G.; Lashgari, A.; Golzani, M. New Inorganic-based Nanohybrids of Layered Zinc Hydroxide/Parkinson's Disease drug and its Chitosan Biopolymer Nanocarriers with Controlled Release Rate *J. Inclu. Pheno.Macro. Chem.* **2017,** *88,* 67–78. DOI: 10.1007/s10847-016-0642-z
2. Hussein, M. Z.; Al Ali, S. H.; Zainal, Z.; Hakim, M. N. l. Development of Antiproliferative Nanohybrid Compound with Controlled Release Property using Ellagic Acid as the Active Agent, *Int. J. Nanomed.* **2011,** *6,* 1373–1383. DOI: 10.2147/IJN.S21567
3. Gu, F. X.; Karnik, R.; Wang, A. Z. Targeted Nanoparticles for Cancer Therapy, *Nano Today,* **2007,** *2,* 14–21. DOI: 10.1016/S1748-0132(07)70083-X
4. Peralta-Videa, P; Zhao, L.; Lopez-Moreno, M.; Dela Rosa, G; Hong, J.; Gardea-Torresdey, Journal ofNanomaterials and the Environment A Review for the Biennium 2008–2010, *J. Haz. Mater.* **2011,** *186,* 1–15. DOI: 10.1016/j.jhazmat.2010
5. Zhang, P.; Webster, T. J. Nanotechnology and Nanomaterials: Promises for Improved Tissue Regeneration, *Nano Today* **2009,** *4,* 66–80. DOI: 10.1016/j.nantod.2008
6. Li, P.; Xiao, F.; Amirkhanian, S.; You, Z. H.; Huang, J. Developments of Nano Materials and Technologies on Asphalt Materials—A Review, *Construc. Build. Mater.* **2017,** *143,* 633–648. DOI: 10.1016/j.conbuildmat.2017
7. Kameliya, J.; Verma, A.; Dutta, P.; Arora, C. H.; Vyas, S. H.; Varma, R. Layered Double Hydroxide Materials: A Review on Their Preparation, Characterization, and Applications, *Inorganics* **2023,** *11,* 1–22. DOI: 10.3390/inorganics11030121
8. Shah, A.; Aftab, S.; Nisar, J.; Ashiq, M. Nanocarriers for Targeted Drug Delivery, *J. Drug Del. Sci. Technol.* **2021,** *62,* 102426. DOI: 10.1016/j.jddst.2021.102426
9. Fu, J.; Jiao, J.; Song, H.; Gu, Z. H.; Liu, Y.; Geng, J.; Jack, K.; Du, A.; Tang, J.; Yu, Ch. Fractal-in-a-Sphere: Confined Self-Assembly of Fractal Silica Nanoparticles, *Chem. Mater*. **2020,** *32,* 314–347. DOI: 10.1021/acs.chemmater.9b03864
10. Putri, O.; Petchsuk, A.; Bayram, S.; Opaprakasit, P. Ultrasonicate-assisted Preparation of Eumelanin-loaded Nano/microparticles based on Polylactide Stereocomplex, *Mater. Today* **2022,** *66,* 3025–3030. DOI: 10.1016/j.matpr.2022.07.08
11. Cozma, D.; Gherca, D.; Mihalcea, I.; Virlan, C.; Cornei, N.; Pui, A. Correlation between Size of CoFe2O4 Nanoparticles Determined from Experimental and Calculated Data

by Different Mathematical Models, *Curr. Nanosci.* **2014,** 10, 869–876. DOI: 10.2174/1573413710666140809010251

12. Ouf, F. X. Sillon, Charging Efficiency of the Electrical Low Pressure Impactor's Corona Charger: Influence of the Fractal Morphology of Nanoparticle Aggregates and Uncertainty Analysis of Experimental Results, Aerosol *Sci. Technol.* **2009,** *43,* 685–698. *DOI:* 10.1080/02786820902878245
13. Lashgari, A.; Ghamami, S.; Bahrami, Z.; Shomossi, F.; Morán, G. S.; Mitnik, D. G. Morphological Investigation and Fractal Properties of Realgar Nanoparticles, *J. Nanomater.* **2015,** 130698. DOI: 10.1155/2015/130698

CHAPTER 13

Impact of Fullerenes and Carbon Nanotubes on the Resistivity of LDPE and i-PP Nanocomposites to the Thermo-Oxidative Exposure

NURLANA MIRZOEVA[1], AIDA DUNYAMALIYEVA[1], NUSHABA KURBANOVA[1], ASGAR HUSEYNOV[2], and ELDAR ZEYNALOV[2]

[1]*Institute of Polymer Materials of the Ministry of Science and Education of the Republic of Azerbaijan, Sumgayit, Azerbaijan*

[2]*Institute of Catalysis and Inorganic Chemistry of the Ministry of Science and Education of the Republic of Azerbaijan, Baku, Azerbaijan*

ABSTRACT

This article presents a perspective overview on the functional thermal properties of polyolefin composites containing nanocarbon additives. Investigations in this perspective field do not still exhaust to bring new interesting results. Thus, a stability of polyethylene and isotactic polypropylene nanocomposites containing fullerene C60/C70 (85/15) and multiwalled carbon nanotubes (MWCNTs) against thermo-oxy onset expansion has been studied. The fullerenes and carbon nanotubes have a high electron affinity (~ 3 eV) and are able to efficiently terminate chain oxidation and degradation processes rising in a polymeric matrix. Along with this feature, the electron microscopy analysis shows uniform distribution of carbon nanoparticles along the nanocomposites that is very important to enable the needed experimental output. A three-dimensional

Advances in Engineering Materials: Preparation for Sustainable Process Development. Tamara Tatrishvili and A. K. Haghi (Eds.)

fine-crystalline supramolecular structure is formed which is necessary prerequisites to have higher thermal stability of nanocomposites. Small amounts of nanofillers introduced into the polymer play the role of structure formers—artificial crystallization nuclei, which contributes to the appearance of a fine spherulite structure in the polymer, characterized by the improved stability states. Indeed, the higher thermo-oxidative stability of nanocomposites was fixed allowing to make purposeful recommendations.

13.1 INTRODUCTION

Nanocarbon compounds are an active component of a package of additives used to improve the properties of polymer materials. Fullerenes, carbon nanotubes and fibers, grapheme, and its derivatives give polymers increased electrical and thermal conductivity, stabilize the polymer matrix against aggressive external factors; they exhibit sensory activity toward many gases.[1,2] The reason for this beneficial effect on the polymer matrix is the outstanding and unique properties of nanocarbon compounds.

First, the electronic structure of molecules (in the case of fullerenes), cylindrical frameworks of nanotubes and fibers, and elegant films of graphenes. Repeatedly condensed aromatic five-six-membered rings generate a highly delocalized electron cloud and acquire an incredible ability to attach nucleophilic particles—the electron affinity of nanocarbon compounds goes beyond 3 Ev.[3] According to numerous experimental facts, fullerenes and other nanocarbon compounds are active scavengers of free radicals and effectively break the destructive chain reactions of oxidation and destruction that develop during the operation of polymer materials under extreme conditions.[4–11] Among the publications on this topic, one can note the increase in the antioxidant efficiency of C_{60} fullerene by combining it with macroperoxide radical scavengers,[12,13] converting C_{60} into water-soluble forms for use in biochemical polymers,[14] and using more effective higher C_{70}–C_{84} fullerenes.[15] Works on the use of C_{60} fullerenes in natural rubber are presented. It was shown in this work that, under conditions of a relatively rarefied polymer structure, C_{60} works both as a thermal stabilizer and as an antioxidant. At the same time, the antioxidant effect begins to decrease with increasing oxygen concentration, which once again indicates inertness to the addition of macroperoxide radicals.[16]

The property of exclusively and actively attaching carbon-centered radicals is most pronounced in carbon nanocompounds with a closed structure, which was emphasized in the example of the thermal stabilization of isotactic polypropylene[17] and the initiated oxidation of 1,4-dioxane and styrene.[18]

The stabilization of polymer materials with carbon nanotubes (CNTs) differs in its mechanism of action from fullerenes and, in this sense, stands apart. The fact is that, despite the high values of electron affinity, their behavior in the polymer matrix is far from unambiguous. As it turned out, the key point is the degree of uniform distribution of CNTs in the polymer matrix. The final result and the effect of CNTs on the polymer depend on this.

Carbon nanostructures have high electrical conductivity and, in the case of a rational uniform distribution in the polymer matrix, they transform dielectrics into materials with high conductivity. On the other hand, high current densities when passing through a polymer material cause strong local heating, leading to the destruction of the material.[19] This means that nanotubes must also have a stabilizing effect, which is the case, making these carbon nanocompounds additives of multifunctional action.

In most publications aimed at improving the physical-mechanical properties and stability of the polymer matrix, CNTs are presented as fillers of polymer materials, for example, when used in PE compositions and polyamides,[20–26] as well as in rubber materials.[27]

In conclusion, the general scheme of reactions that ensure effective stabilization of polymer materials can be presented as follows.[28–30]

The purpose of this work is to study the effect of the carbon nanoparticles on the thermal properties of nanocomposites based on low-density high-pressure PE and isotactic PP.

13.2 EXPERIMENTAL PART

In the work were used: high-pressure polyethylene (LDPE), grade 15803-020 from SOCAR, (Azerbaijan); isotactic polypropylene (PP) 01030 brand "Kaplen" (Russia), melt flow index (MFI) 2.3÷3.6 g • 10^{-1} min, molecular weight ~ 2–3 × 10^5, polydispersity index 4.5; a mixture of fullerenes $C_{60/C70}$ (85/15 wt.%) was provided by Aldrich (USA).

Nanocomposite polymer materials were prepared by mixing LDPE and PP in the presence of a mixture of $C_{60/C70}$ fullerenes and MWCNTs on laboratory rollers at a temperature of 160–165°C for 15 minutes.

The thermal stability of the nanocomposite samples under study was studied using a Q-1500D derivatograph from MOM, Hungary. The tests were carried out in an air atmosphere in a dynamic mode when the sample was heated at 5 deg/min^{-1} from 20 to 500°C, the test portion of 100 mg, sensitivity of the channels DTA-250 μV, TG-100, DTG-1 mV.

The thermal stability of the studied samples based on LDPE -MWCNTs was assessed by the activation energy (E_a) of thermal oxidative destruction, calculated by the double logarithm method according to the TG curve by the temperature of 10% (T_{10}), 20% (T_{20}), and 50% (T_{50}) decay of the samples under study, as well as their half-life period $-t_{1/2}$.

13.3 RESULTS AND DISCUSSION

The results of investigations on thermo-oxidative resistance of aforementioned nanocomposites are presented in Tables 13.1 and 13.2.

TABLE 13.1 Thermal Properties of the i-PP-$C_{60/70}$ Nanocomposites.

Composition formulation (wt.%)	T_{10}, %	T_{20}, %	T_{50}, %	$\tau_{1/2}$, min	E_a, kJ/mol
PP(100)	135	265	335	63.2	145.45
PP/$C_{60/70}$(100/0.01)	235	280	345	64.4	186.61
PP/$C_{60/70}$(100/0.02)	250	300	360	68.8	189.48
PP/$C_{60/70}$(100/0.04)	230	250	340	65.6	184.26

TABLE 13.2 Thermal Properties of the LDPE-MWCNTs Nanocomposites.

Composition formulation (wt.%)	T_{10}, %	T_{20}, %	T_{50}, %	$\tau_{1/2}$, min	E_a, kJ/mol
LDPE (100)	320	340	350	63.2	129.45
LDPE/MWCNT (100/0.01)	355	375	405	69.1	148.17
LDPE/MWCNT (100/0.05)	360	380	410	72.3	165,22
LDPE/MWCNT (100/0.1)	350	370	400	68.6	146.84

It can be seen that introduction of a mixture of fullerenes $C_{60/70}$ into the PP composition helps to increase the decomposition temperature of the samples: T_{10} from 135 to 250°C, T_{20} from 265 to 300°C, T_{50} from 335 to 360°C, half-life period ($t_{1/2}$) from 63.2 to 68.8 min, activation energy (E_a)

of thermal oxidative destruction of the resulting nanocomposites increases from 145.45 to 189.48 kJ/mol.

The Vicat softening point of the resulting compositions showed that the introduction of a mixture of $C_{60/70}$ fullerenes (0.02 wt%) into the PP composition as a nanofiller leads to an increase in the heat resistance index from 165 to 230°C

The introduction of MWCNTs into LDPE matrix leads to an increase in T_{50} from 350 to 410°C, $t_{1/2}$ from 63.2 to 72.3 min, E_a from 129.45 to 165.22 kJ/mol. The combined use of both NiO NPs and MWCNT in LDPE contributes to an increase in T_{50} to 420°C, $t_{1/2}$ to 79.6, and E_a to 267.91 kJ/mol.

13.4 CONCLUSION

The effect of nickel oxide nanoparticles stabilized by a high-pressure polyethylene matrix obtained by the mechano-chemical method on the properties of high-pressure polyethylene-based composites containing MWCNT has been studied.

An improvement of thermal and oxidative stability of LDPE and PP nanocomposites upon introducing $C_{60/70}$ fullerenes and MWCNTs has been unequivocally proved. The small amounts of carbon nanocompounds operate in polymeric matrix as power radical scavengers and efficient formers of uniform structure as well.

KEYWORDS

- **polyethylene nanocomposites**
- **polypropylene nanocomposites**
- **fullerenes**
- **multi-walled carbon nanotubes**
- **electron microscopy**
- **differential-thermal analysis**
- **thermogravimetric analysis**

REFERENCES

1. Kumar, A. P.; Depan, D.; Tomer, N. S.; Singh, R. P. Nanoscale Particles for Polymer Degradation and Stabilization—Trends and Future Perspectives. *Prog. Polym. Sci.* **2009,** *34*, no. 6, 479–515.
2. Jineesh, A. G.; Mohapatra, S. Thermal Properties of Polymer–Carbon Nanocomposites. In *Carbon-Containing Polymer Composites*; Springer: Singapore, 2019; pp 235–270.
3. Cioslowski, J. *Electronic Structure Calculations on Fullerenes and Their Derivatives*, Oxford University Press, 1995.
4. Krusic, P. J.; Wasserman, E.; Keizer, P. N.; Morton, J. R.; Preston, K. F. Radical Reactions of C_{60}. *Science* **1991,** *254*, no. 5035, 1183–1185.
5. Krusic, P. J.; Wasserman, E.; Parkinson, B. A.; Malone, B.; Holler, Jr, E. R.; Keizer, P. N.; Preston, K. F. Electron Spin Resonance Study of the Radical Reactivity of C_{60}. *J. Am. Chem. Soc.* **1991,** *113*, no. 16, 6274–6275.
6. McEwen, C. N.; McKay, R. G.; Larsen, B. S. C_{60} as a Radical Sponge. *J. Am. Chem. Soc.* **1992,** *114*, no. 11, 4412–4414.
7. Walbiner, M.; Fischer, H. Rate Constants for the Addition of the Benzyl Radical to Fullerene C_{60} in Solution. *J. Phys. Chem.* **1993,** *97*, no. 19, 4880–4881.
8. Zeynalov, E. B.; Allen, N. S.; Salmanova, N. I. Radical Scavenging Efficiency of Different Fullerenes C_{60}, C_{70} and Fullerene Soot. *Polym. Degrad. Stab.* **2009,** *94*, no. 8, 1183–1189.
9. Dimitrijevic, N. M.; Kamat, P. V.; Fessenden, R. W. Radical Adducts of Fullerenes C_{60} and C_{70} Studied by Laser Flash Photolysis and Pulse Radiolysis. *J. Phys. Chem.* **1993,** *97*, no. 3, 615–618.
10. Zeinalov, E. B.; Koβmehl G. Fullerene C_{60} as an Antioxidant for Polymers. *Polym. Degrad. Stab.* **2001,** *71*, no. 2, 197–202.
11. Zeynalov, E. B.; Magerramova, M.; Ishenko, N.; Fullerenes C_{60}/C_{70} and C_{70} as Antioxidants for Polystyrene. *Iran. Polym. J.* **2004,** *13*, no. 2, 143–148.
12. Troitskii, B. B.;Troitskaya, L. S.; Dmitriev, A. A.; Anikina, L. I.; Novikova, M. A.; Denisova, V. N.; Khokhlova, L. V. Fullerenes and Synergistic Mixtures on their Basis as High-Temperature Antioxidants of Polymers. *Mol. Mater.* **2000,** *13*, nos. 1/4, 209–212.
13. Nekoei, A. R.; Haghgoo, S. DFT Investigation on Some Nitrogen-Doped Fullerenes with More Antiradical and Antioxidant Activities than C_{60}. *Struct. Chem.* **2019,** *2019*, no. 5, 1–12.
14. Tam, J.; Liu, J.; Yao, Z. Effect of Microstructure on the Antioxidant Properties of Fullerene Polymer Solutions. *RSC Adv.* **2013,** *3*, no. 14, 4622–4627.
15. Shibaev, L. A.; Egorov, V. M.; Zgonnik, V. N.; Antonova, T. A.;Vinogradova, L. V.; Melenevskaya, E. Y.; Bershtein, V. A. An Enhanced Thermal Stability of Poly (2, 6-Dimethyl-1,4-Phenylene Oxide) in the Presence of Small Additives of C_{60} and C_{70} Fullerenes. *Polym. Sci. Ser. A* **2001,** *43*, no. 2, 101–105.
16. Cataldo, F. On the Reactivity of C_{60} Fullerene with Diene Rubber Macroradicals. I. the Case of Natural and Synthetic Cis-1, 4-Polyisoprene Under Anaerobic and Thermooxidative Degradation Conditions. *Fullerene Sci. Technol.* **2001,** *9*, no. 4, 497–513.
17. Jipa, S.; Zaharescu, T.; Santos, C.; Gigante, B.; Setnescu, R.; Setnescu, T.; Olteanu, R. L. The Antioxidant Effect of Some Carbon Materials in Polypropylene. *Mater. Plast.* **2002,** *39*, no. 1, 67–72.

18. Yumagulova, R. K.; Medvedeva, N. A.;Yakupova, L. R.; Kolesov, S. V.; Safiullin, R. L. Free-Radical Chain Oxidation of 1,4-Dioxane and Styrene in the Presence of Fullerene C_{60}. *Kinet. Catal.* **2013,** *54*, no. 6, 709–715.
19. Watts, P. C. P.; Fearon, P. K.; Hsu, W. K.; Billingham, N. C.; Kroto, H. W.; Walton, D. R. M. Carbon Nanotubes as Polymer Antioxidants. *J. Mater. Chem.* **2003,** *13*, no. 3, 491–495.
20. Sreekanth, P. R.; Kumar, N. N.; Kanagaraj, S. Improving Post Irradiation Stability of High Density Polyethylene by Multi Walled Carbon Nanotubes. *Compos. Sci. Technol.* **2012,** *72*, no. 3, 390–396.
21. Shi, X.; Jiang, B.; Wang, J.; Yang, Y. Influence of Wall Number and Surface Functionalization of Carbon Nanotubes on their Antioxidant Behavior in High Density Polyethylene. *Carbon* **2012,** *50*, no. 3, 1005–1013.
22. Bocchini, S.; Frache, A.; Camino, G.; Claes, M. Polyethylene Thermal Oxidative Stabilisation in Carbon Nanotubes Based Nanocomposites. *Eur. Polym. J.* **2007,** *43*, no. 8, 3222–3235.
23. Shen, Z.; Bateman, S.; Wu, D. Y.; McMahon, P.; Dell'Olio, M.; Gotama, J. The Effects of Carbon Nanotubes on Mechanical and Thermal Properties of Woven Glass Fibre Reinforced Polyamide-6 Nanocomposites. *Compos. Sci. Technol.* **2009,** *69*, no. 2, 239–244.
24. Gorrasi, G.; Sarno, M.; Di Bartolomeo, A.; Sannino, D.; Ciambelli, P.; Vittoria, V. Incorporation of Carbon Nanotubes Into Polyethylene by High Energy Ball Milling: Morphology and Physical Properties. *J. Polym. Sci. Part B* **2007,** *45*, no. 5, 597–606.
25. Martinez-Morlanes, M. J.; Castell, P.; Alonso, P. J.; Martinez, M. T.; Puertolas, J. A. Multi-Walled Carbon Nanotubes Acting as Free Radical Scavengers in Gamma-Irradiated Ultrahigh Molecular Weight Polyethylene Composites. *Carbon* **2012,** *50*, no. 7, 2442–2452.
26. Guadagno, L.; Naddeo, C.; Raimondo, M.; Gorrasi, G.; Vittoria, V. Effect of Carbon Nanotubes on the Photo-Oxidative Durability of Syndiotactic Polypropylene. *Polym. Degrad. Stab.* **2010,** *95*, no. 9, 1614–1626.
27. Kodal, M.; Karakaya, N.; Wis, A. A.; Ozkoc, G. Thermal Properties (DSC, TMA, TGA, DTA) of Rubber Nanocomposites Containing Carbon Nanofillers. In *Carbon-Based Nanofiller and Their Rubber Nanocomposites*, 2019; pp 325–366.
28. Zeynalov, E. B.; Agaguseynova, M. M.; Salmanova, N. I. Effect of Nanocarbon Additives on Stability of Polymer Composites. *Izv. Vyssh. Uchebn. Zaved. Khim. Khim. Tekhnol.* [Russ. J. Chem. & Chem. Tech.]. **2020,** *63*, no. 11, 4–11.
29. Dunyamaliyeva, A. I.; Kurbanova, N. I.; Zeynalov, E. B. Fullerene-Containing Composites Based on Isotactic Polypropylene. *Azerbaijan Chem. J.* **2021,** *2*, 24–28.
30. Zeynalov, E. B.; Allen, N. S.; Huseynov, A. B.; Kurbanova, N. I.; Mirzoyeva, N. A.; Dunyamaliyeva, A. I. Overview on MWCNT Polymeric Composites and Novel Properties of MWCNT HDPE Composite Containing Nanoparticle Cobalt Oxide. *Curr. Trends Polym. Sci.* **2021,** *21*, 17–25.

CHAPTER 14

Zinc-Containing Thermoplastic Elastomer Nanocomposites on the Basis of Isotactic Polypropylene

TURKAN GULIYEVA, NUSHABA KURBANOVA, SEVINC RAGIMOVA, ISMAIL ISMAILOV, and DILBAR NURULLAYEVA

Institute of Polymer Materials of the Ministry of Science and Education of the Republic of Azerbaijan, Sumgait, Azerbaijan

ABSTRACT

New nanocomposites based on isotactic polypropylene (PP), butadiene-nitrile rubber (BNR), and nanofillers (NF) have been prepared. The effect of the zinc oxide nanoparticles, stabilized in a high-pressure polyethylene matrix (PE), concentration of the properties of the nanocomposites is studied. X-ray phase (XPhA) and differential thermal (DTA) analyses are applied for investigation into the structure and properties of the obtained nanocomposite materials. It is demonstrated that the obtained zinc-containing nanocomposites have high thermal stability. The physical-mechanical properties of nanocomposites based on PP/BNR and zinc oxide nanoparticles have also been investigated. It has shown that the introduction of zinc oxide nanoparticles in PP/BNR increases its strength, deformation, and rheological parameters.

Advances in Engineering Materials: Preparation for Sustainable Process Development. Tamara Tatrishvili and A. K. Haghi (Eds.)

14.1 INTRODUCTION

The development of the production of polymer composition materials and the continuous expansion of their application areas has been closely connected with the successes in the field of creation of nanocomposites.

Now, one of the polymer materials with a constantly growing share of world consumption are thermoplastic elastomers (TPE). In various branches of industry, TPE based on binary mixtures of polyolefin and rubber are widely used, and the investigations are also expanded on the preparation of materials on the basis of various initial components, which allows changing the properties of thermoplastic elastomer materials for meeting a variety of requirements.[1–6]

However, at mixing a polar elastomer phase and a nonpolar thermoplastic phase in a BNR—PP pair, poor polymer compatibility occurs, resulting in the material having low physical-mechanical and thermal characteristics. One of the possible ways of improvement of the complex of properties of compositions based on BNR—PP is the introduction of nano-sized fillers (NF), being an interphase additive, contributing to the improvement of both the combination of components and the exploitation properties of the resulting substances into its composition.[7–12] In this regard, it is interesting and important to study the influence of nano-sized metal-containing particles on the properties of TPE characterizing their structure, and to establish the availability of a correlation of the latter ones with physical-mechanical, thermal, and operational properties.

However, as we all know, by adding d- valence (nickel, cobalt, copper, etc.) metal nanoparticles (NP) to the polymer, new types of nanocomposites are obtained, which can be applied to optics, medicine, optoelectronics, and many other fields.[13–15]

In this paper, the influence of small additives of NF containing NP of zinc oxides on the physical-mechanical parameters and thermal stability of the obtained compositions based on isotactic PP and BNR has been investigated.

14.2 EXPERIMENTAL METHODS AND MATERIALS

The following polymers were used as a polymer matrix:

- isotactic PP "Kaplen" (Russia) of mark 01 030 with molecular weight $\sim 2\text{–}3 \times 10^5$, heterogeneity index 4.5, MFI 2.3–3.6 $g \cdot 10^{-1}$ min.

- butadiene–nitrile copolymer (BNR) of mark NB 192 HF, containing 27% of acrylonitrile, firm BSL Olefinverbund GmbH Shckopau, d=0.98г/см^3 (Germany).

The nanoparticles (NP) of zinc oxide (ZnO) stabilized by a polymer matrix of high-pressure polyethylene, obtained by a mechano-chemical method, were used as nanofiller. Nanoparticle content—5 mass %, size 39±1.0 nm, degree of crystallinity—35 – 45%.[16–18] Ratio of the components of the composition (mass.%): PP/BNR/NF=50/50/(0.5; 1.0; 2.0).

Metal-containing polymer materials have been obtained by mixing PP with BNR and zinc–containing NF on laboratory rollers at temperature 160–165°C for 15 min. For carrying out of the mechanical testing, the prepared nanocomposites were pressed with thickness of 1 mm at 190°C and pressure 10 MPa.

The physical-mechanical properties of the prepared nanocomposites were defined on the device RMI-250.

X-ray phase analysis (XPhA) of the obtained compositions was carried out on the device "D2 Phaser" by Bruker firm (Germany).

The melt flow index (MFI) has been determined on the device IIRT at temperature equal to 230°C, load—5.0 kg.

The heat resistance of the polymer nanocomposites was determined by the Vicat method.

The thermal stability of the investigated samples of thermoelastoplastics was studied on a derivatograph of mark Q-1500D of the association MOM (Hungary). The experiments were carried out in room conditions in dynamic mode when the sample was heated 5 grad·min^{-1} from 20 to 500°C, the weight was 100 mg, DTA-250mcV, TG-100, DTG-1mV.

Experiments were carried out in room conditions in dynamic mode when the sample was heated.

14.3 RESULTS AND DISCUSSION

First of all of the investigation, the samples of thermoelastoplastics based on mixtures of polypropylene and butadiene–nitrile rubber with a zinc-containing nanofiller have been made. The proportion of the primary components (mass%): PP/BNR/NF=50/50/(0.5; 1.0; 2.0). The phase composition and physical-mechanical, heat-physical parameters, and thermo oxidative stability of the prepared nanocomposites have been studied.

The influence of zinc oxide nanoparticles on the structure of PP/BNR compositions has been investigated by the X-ray phase method. The phase explanation was performed in concordance with data on interplanar spacing, using the ASTM card file. In Figure 14.1, it has been presented a diffractogram of PP/BNR sample, on which reflexes corresponding to PP are visible: d_{hkl}, holoimage is determined for a formless butadiene–nitrile rubber. On the diffractogram of PP/BNR/NF sample (Figure 14.2) the reflexes characteristic for zinc-containing nanoparticles is also observed: d_{hkl} 2.47966; 2.13735; 1.51401; 1.28859 Å (Figure 14.2), which corresponds to d_{hkl} series of zinc oxide according to ASTM file.

In Table 14.1, the results of physical and mechanical studies of samples of zinc-containing TPE nanocomposite on the basis of PP/NBR are presented.

From the data in Table 14.1 it can be clearly seen that the strength index increases from 5.04 to 6.12 MPa when a nanofiller containing 1.0 wt.% zinc oxide is included in the composition of the nanocomposite. An increase of NF concentration from 1.0 to 2.0 mass% leads to causes a decrease in the strength of the PP/BNK/NF mixture (5.53 MPa), which has been probably connected with the aggregation of nanoparticles, leading to the created offlaws in a volume of the polymer. The addition of 1.0 mass % of NN into the composition leads to an increase of value of specific elongation of the nanocomposite by 1.75÷2.25 times, which has been apparently stipulated by the synergetic effect connected with the interphase interaction of zinc-containing nanoparticles in PE matrix with the components of PP/BNR composition, the mutual influence of which favors the increase in both the deformation value and the strength index. Looking at the heat resistance results according to Vicat, it can be seen that adding nanofillers to the composition of PP/BNK leads to an increase in the heat resistance index from 87 to 129°C. In this case, the introduction of 0.5 mass % NF leads to an increase in the heat resistance index to 129°C, a further increase of a quantity of NF leads to a decrease in the heat resistance index, which has probably been stipulated by the microdeficiency of the obtained composite.

Comparing from the data presented in this table, with a further raise of nanofiller content, the growth of tensile yield strength is observed. Improving the melt flow index expands the processing areas of the resulting nanocomposites, since they can be processed by injection molding and extrusion methods.

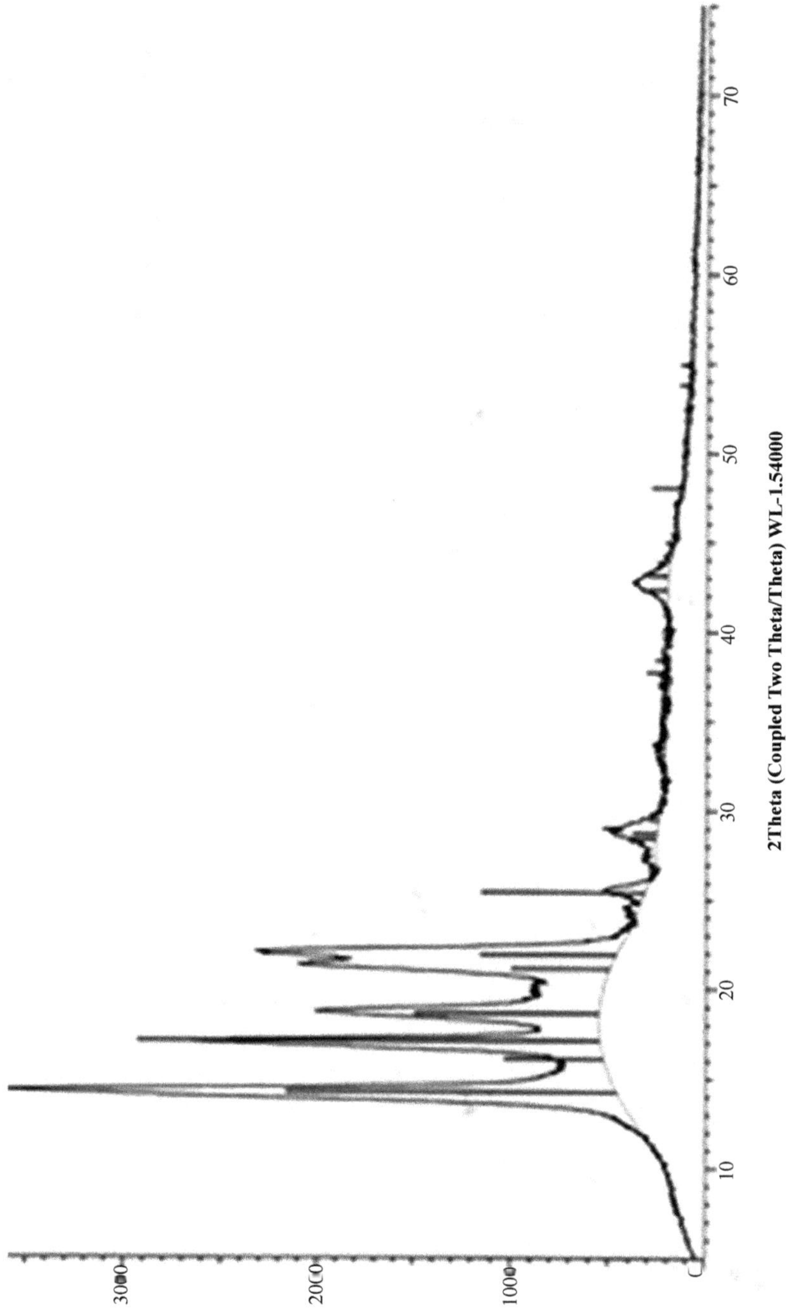

FIGURE 14.1 Diffractogram of PP/BNR sample.

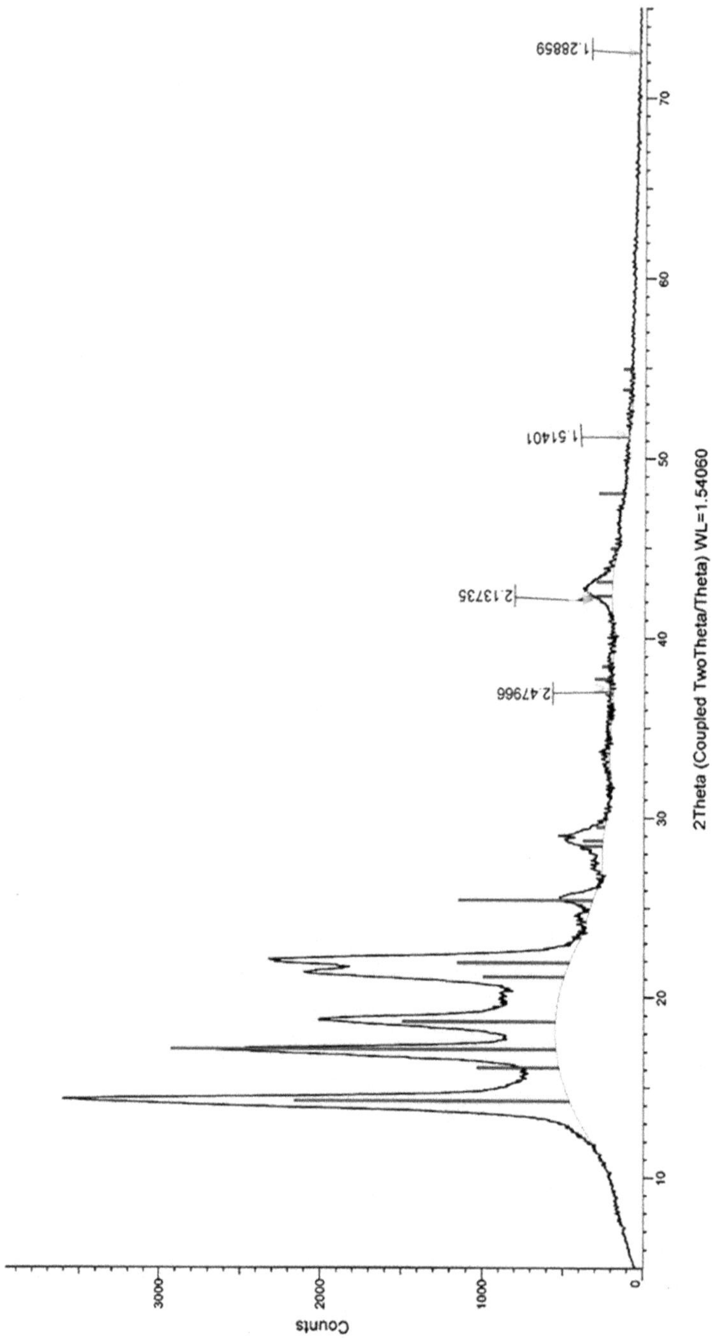

FIGURE 14.2 Diffractogram of PP/BNR/NF sample.

TABLE 14.1 Physical-Mechanical and Heat-Physical Parameters of Composition Materials.

S. No.	Composition (mass %.), PP/BNR /NF	Tensile strength, Mpa	Specific elongation, %	Vicat heat-stability,°C	Melt flow index (MFI), g/10 min.
1	50/50/0	5.04	16	87	0.089
2	50/50/0.5	5.90	28	124	0.108
3	50/50/1.0	6.12	36	129	0.118
4	50/50/2.0	5.53	32	121	0.146

The thermal durability of the investigated samples of mixed TPE on the basis of PP/BNR and PP/BNR/NF containing NF with NP of zinc oxide was estimated on the activation energy (E_a) of the decay of thermal-oxidative degradation determined on TG curve according to the methodology (Kryzhanovsky, 2007), on the decay T_{10}, T_{20}, T_{50} (loss of weight 10%, 20%, 50%) of the investigated TPE samples, and also on their half-decay time—$\tau_{1/2}$.

The results of derivatographic investigations are presented in Table 14.2.

TABLE 14.2 Thermal durability of the Investigated Samples of Nanocomposites.

Composition (mass %.) PP/BNR /NF	M.p.,°C	T_{10},°C	T_{20},°C	T_{50},°C	$\tau_{1/2}$, min.	E_a, J·mol^{-1}
50/50/0	150	225	250	300	62.8	124.48
50/50/0.5	150	230	285	355	64.9	162.14
50/50/1.0	150	245	320	370	70.6	172.37
50/50/2.0	150	235	290	360	66.7	164.53

According to the information given in the table, the introduction of zinc oxide nanoparticles into the composition of mixed thermoelastoplastics favors the considerable increase in the temperature of the decay of the samples: T_{10} by 10–20°C, T_{20} by 35–70°C, T_{50} by 55–70°C; half-decay time $\tau_{1/2}$ is increased from 62.8 to 70.6 min, and activation energy (E_a) of decay of thermal-oxidative destruction of the obtained nanocomposited is increased from 124.48 to 172.37 kJ/mol. The introduction of a nanofiller into the composition of PP/BNR practically does not influence the melting point (m.p.) of the composition.

The results of derivatographic analysis showed that the addition of NF containing NP of zinc oxide into the composition of mixed thermoplastics favors the improvement of the thermal properties of the compositions based on isotactic PP and BNR nanocomposites.

The permolecular structure of the composites (rate of crystallinity, the availability of various functional groups and different branches, etc.) and the interphase interaction at the interface noticeably influence the properties of polymer composites.[20,21]

The metal-containing nanoparticles used in the work, located at the boundary between phase stratum of the component of PP, BNR, and PE, favor the formation of heterogeneous nucleation centers in the melt of composition, which in the stepwise cooling process of the nanocomposite favor the increase in the crystallization centers. As a result, the crystallization process of the nanocomposite improves and a fine relatively spherulite structure is formed.

14.4 CONCLUSION

The influence of a nanofiller containing zinc oxide nanoparticles on the peculiarities of properties of mixed TPE on the basis of isotactic PP and BNR has been investigated. XPhA diffractograms prove that PP/BNK/NF nanocomposites contain zinc oxide nanoparticles.

It has been revealed the improvement of strength, deformation and rheological parameters, as well as the thermal-oxidative stability of the obtained nanocomposites, which has been apparently connected with the synergetic effect of the interaction of zinc-containing nanoparticles in PE matrix with the components of PP/BNR polymer composition.

It has been shown that the mixed TPEs on the basis of PP/BNR can be processed only by a method of pressing and on the basis of PP/BNR/NF—both by pressing and by injection molding and extrusion methods.

It has been shown the perspectivity of use of nanofiller containing NP of zinc oxide, an additive to PP/BNR polymer composition, which the nanofiller causes the formation of a fine spherulitic structure in the nanocomposite, thereby improving its properties and the application areas of the resulting nanocomposite are expanded.

KEYWORDS

- **isotactic propylene**
- **butadiene-nitrile rubber**
- **nanoparticles of zinc oxide**
- **physical-mechanical and thermal properties**

REFERENCES

1. Wolfson, S. I.; Okhotina, N. A.; Panfilova, O. A.; Novikova, E. V.; Minnegaliev, R. R. Dynamically Vulcanized Thermoplastic Elastomers Based on a Mixture of Rubbers of Different Polarities and Polypropylene. *Vestn. Kazan. Technol. Univ.* **2015,** *18*, 90–92.
2. Polonik, V. D.; Prokopchuk, N. R.; Shashok, Z. S. Technical Properties of Elastomer Compositions on the Basis of Butadiyen-Nitril of Rubber Modified by Politetraftoretilen. *Chem. Technol. Organ. Substances Mater. Goods* **2012,** 85–87.
3. Okhotina, N. A.; Nigmatullina, A. I.; Sabirov, R. K. Study of Elastic-Hysteresis Characteristics of Dynamic Thermoplastic Elastomers. *Bull. Kazan Technol. Univ.* **2012,** 100–101.
4. Stepanov, A. V.; Zaderenko, T. V.; Goncharuk, G. P., *et al.* Structure and Properties of Mixtures Based on Polyethylene and Ethylene Propylene Diene Rubber. *Rubber Resin* **2010,** *3*, 8–11.
5. Zaikin, A. E.; Bobrov, G. B. Compatibilization of Blends of Incompatible Polymers Via Filling. *Polym. Sci. Ser. A* **2012,** *54*, 651–657.
6. Allahverdiyeva, K. V.; Kakhramanov, N. T.; Martynova, G. S.; Mustafayeva, F. A.; Gahramanli, Y. N.; Habibova, A. G.; Bafadarova, H. B.; Suleymanova, E. I.; Gurbanova, R. V. Structural Features and Mechanism of Crystallization of Nanocomposites Based on Maleinated High Density Polyethylene and Carbon Black. *Heliyon* **2023,** doi: 10.1016/j.heliyon.2023.e14829.
7. Wolfson, S. I.; Okhotina, N. A.; Nigmatullina, A. I., *et al.* Elastic-Hysteresis Properties of Dynamic Thermoplastic Elastomers Modified With Nanofiller. *Plastics* **2012,** *4*, 42–45.
8. Panfilova, O. A.; Wolfson, S. I.; Okhotina, N. A.; *et al.* Compatible Additives for Enhancing Interaction at the Phase Boundary in Thermoplastic Vulcanizates Based on Rubbers of Different polarity and Polypropylene. *Rubber Resin* **2017,** *4*, 224–229.
9. Kakhramanov, N. T.; Allahverdiyeva, K. V.; Gahramanli, Y. N.; Mustafayeva, F. A. Physical–Mechanical Properties of Multifunctional Thermoplastic Elastomers Based on Polyolefins and Styrene-Butadiene Elastomer. *J. Elastomers Plastics* **2023,** *55*, 279–302.

10. Perestoronina, Z. A.; Ableev, R. I.; Baranets, I. V.; Kurlyand, S. K. Effect of Polymer Additives on Enhancing Interfacial Interaction in Blended Thermoplastic Elastomers. *Rubber* **2012,** 13–16.
11. Kuchmenova, L. H.; Slonov, A. L.; Zhansitov, A. A.; *et al.* Study of Thermal Properties of Polymer-Polymer Composites Based on Polypropylene. *Plastics* **2014,** *7 8*, 7 9.
12. Нигматулина, А. И.; Вольфсон, С. И.; Охотина, Н. А.; Шалдыбина, М. С. Свойства динамических термоэластопластов содержащих модифицированных полипропилен и слоистый наполнитель. *Вестн. Казан. технол. Ун-та.* **2010,** *9,* 329–331.
13. Kurbanova, N. I.; Guliyeva, T. M.; Ishenko, N. Y. Preparation and Properties of Metal- Containing Nanocomposites Based on Isotactic Polyprophlene and Butadiene-Acrylonitrile Rubber. *Russ. J. Appl. Chem.* **2021,** *94*, 17–21.
14. Kurbanova, N. I.; Ragimova, S. K.; Alimirzoeva, N. A.; Ishenko, N. Y.; Medyakov, V. V.; *et.al.* Composites Based on Isotactic Polypropylene and High-Pressure Polyethylene with Zinc-Containing Nanofillers, *Inorgan. Mater. Appl. Res.* **2022,** *13*, 485–488.
15. Alimirzayeva, N. A. The Effect of Metal-Containing Nanoparticles on the Thermo—Oxidative Properties of Thermoplastic Elastomers Based on Isotactic Polypropylene and Ethylene Propylene Diene Rubber. *New Mater. Compounds Appl.* **2022,** *6*, 220–229.
16. Kurbanova, N. I.; Alimirzoeva, N. A.; Guseinova, Z. N.; Nurullayeva, D. R. *Ecological Method of Preparation of Metal-Containing Nanoparticles in Polyethylene Matrix*, ITWCCST, Book of Proceedings, Baku, Azerbaijan, 10–13 Sept, 2017; pp. 24–26.
17. Kurbanova, N. I.; Kuliyev, A. M.; Alimirzoeva, N. A.; Aliyev, A. T.; Ishenko, N. Y.; Nurullayeva, D. R. Preparation of Copper-Containing Nanoparticles in Polyethylene Matrix Without Use of Solvents. In *Science and Technology of Polymers and Advanced Materials: Applied Research Methods;* Mukbanıanı, O. V., Tatrıshvılı, T. N., Abadie to be published by Apple Academic Press, Inc., 2019; pp 57–65.
18. Ragimova, S. K.; Obtaining the Metal-Containing Nanoparticles in Polyethylene Matrix by Mechano-Chemical Method and Study of their Properties. *Az. Chem. J.* **2020,** *2*, 20–25.
19. Kryzhanovsky, V. K. *Technical Properties of Polymeric Materials: Textbook and Reference Manual*, Profession, St. Petersburg, 2007, 240 c.
20. Kuleznev, V. N. *Polymer Mixtures and Alloys*. Lecture notes, Scientific Foundations and Technologies: St. Petersburg, 2013, 216 c.
21. Pomogailo, A. D. Molecular Polymer-Polymer Compositions. Synthetic Aspects. *Russ. Chem. Rev.* **2002,** *71*, 1–31.

CHAPTER 15

Thermal Conductivity Properties of MW-CNT-Based Natural Azerbaijani Gobu Bentonite Ceramic Composites

SAMIRA MAMMADOVA[1,2], ELDAR ZEYNALOV[3], and ASGAR HUSEYNOV[3]

[1]*G. M. Abdullayev Institute of Physics, Ministry of Science and Education, Republic of Azerbaijan, Baku, Azerbaijan*

[2]*National Aviation Academy, Baku, Azerbaijan*

[3]*Institute of Catalysis and Inorganic Chemistry Named After Academician M.Naghiyev, Ministry of Science and Education, Republic of Azerbaijan, Baku, Azerbaijan*

ABSTRACT

This research work reports the thermal conductivity measurement results of the ceramic composites based on natural Azerbaijani bentonite of Gobu and multi-walled carbon nanotubes. Nanocomposites were prepared in different concentrations of x MW-CNTs (x = 1%; 2%; 4%; 8%; 16%) for comparison of their thermal conductivity depending on concentration. For measuring thermal conductivity properties, we have used a laser flash technique. The results have shown that with increasing concentrations of MW-CNTs, the thermal conductivity is increasing as expected. For example, for 16% MW-CNT-containing composite, the highest thermal conductivity was observed (κ = 1.3 W/(m·K)) at T = 498 K. It was determined that, in spite of the fact that CNTs have a higher thermal

Advances in Engineering Materials: Preparation for Sustainable Process Development. Tamara Tatrishvili and A. K. Haghi (Eds.)

conductivity as nanofillers, inside ceramic matrices it decreases. However, the thermal conductivity of these nanocomposites is not high, but they are very useful for the preparation of coolers in electronic devices as well as heating elements.

15.1 INTRODUCTION

As nanotechnology is constantly evolving, the size of electronic circuits in devices is decreasing to quantum sizes. For such devices, effective coolers should be prepared. Due to their stable resistive properties at high temperatures, ceramics are considered very useful materials for these applications. Because of the wide application demand in the field of microelectronics and thermal insulators, nanocomposites based on ceramics with high and low thermal conductivity are very relevant.[1,2]

Literature review showed that CNTs in a single case have a high thermal conductivity of approximately ~ 2000 – 3000 W/·(m*K).[3–5] Therefore, it is very important to analyze their thermal conductivity properties inside different matrices like polymers, ceramics. It is obvious from the literature that, many authors have used synthetic ceramics as matrices. However, synthetic ceramics are very expensive. Therefore, we have used natural and affordable bentonites as matrix and MW-CNTs as a reinforcement material.

15.2 EXPERIMENTAL METHODS AND MATERIALS

MWCNTs were synthesized using a xylene mixture (~ 98.5%) as a carbon source and ferrocene (98%) as a catalyst by the aerosol-chemical vapor deposition (A-CVD) method in a laboratory setup from Schidre.[6] Ar (99.999%) and H_2 (99.999%) gases were obtained from Farafan Gas Co., Tehran, Iran. Natural bentonite clays were taken from the Gobu village of Azerbaijan.[7] We have used the irreversible dispersion method for the preparation of MWCNT-based natural bentonite ceramic composites. For measuring thermal conductivity, nanocomposites were prepared in a rectangular shape in 10.0 mm × 10.0 mm × 2.9 mm sizes. The prepared samples were calcined at 950C in an Ar atmosphere.

15.3 RESULTS AND DISCUSSION

The morphological structure of the Gobu bentonite and Gobu/ MWCNT (4%) ceramic composite was represented in Figures 15.1a and 15.1b. As shown from Figure 15.1 inside Gobu bentonite, MWCNTs were distributed horizontally to each other.

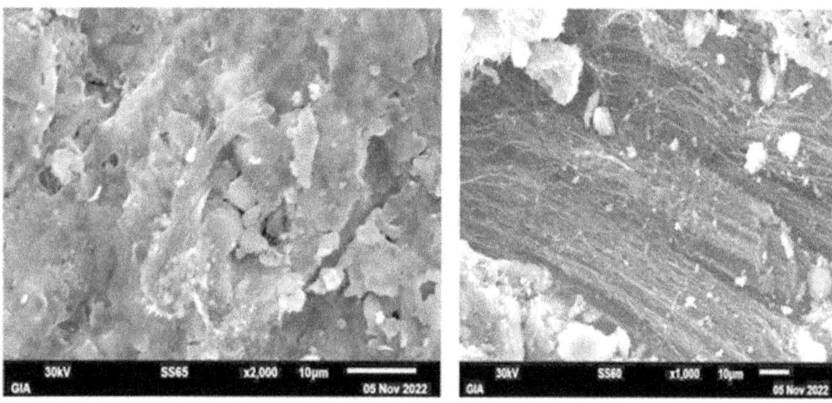

a) b)

FIGURE 15.1 SEM images of (a) Gobu bentonite in Ar atmosphere and (b) Gobu/x MWCNT (x = 4%) ceramic composite at 10 μm scale.

Source: Reprinted with permission from Ref. [11]. Copyright © 2023, Springer Nature

The thermal conductivity properties of the prepared composites were studied depending on temperature. The results show that thermal conductivity of the prepared nanocomposites is not very high, which can be explained with different parameters. For example, the structure, type length, MW-CNT network form, phonon scattering from crossed CNTs, interfaces between CNTs and bentonites. In general, less research is done in the literature on the thermal properties of natural clay bentonite and carbon nanomaterial-based nanocomposites, particularly their thermal conductivity properties.[8,9] Generally, literature review proved our results; thus, a lot of composites based on natural bentonite clays and different types of nanocarbon materials show low thermal conductivity.[10–13] The thermal conductivity results of the Gobu/ x MWCNT are shown in Figure 15.2. Measurements were carried out at a wide temperature range (25–300°C) for different concentrations x MWCNTs (x = 1%; 2%; 4%; 8%; 16%) contained composites.

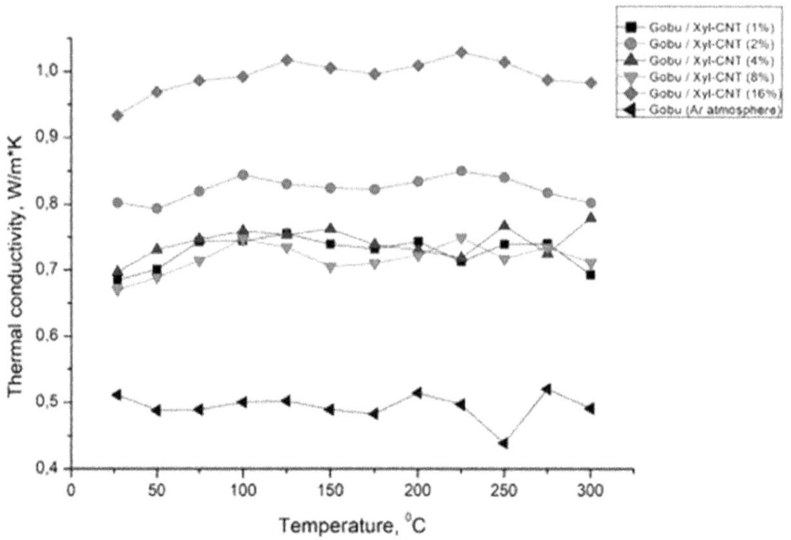

FIGURE 15.2 Thermal conductivity of Gobu/ xMWCNT (x = 1%; 2%; 4%; 8%; 16%) nanocomposites at (300–573 K) temperature ranges.

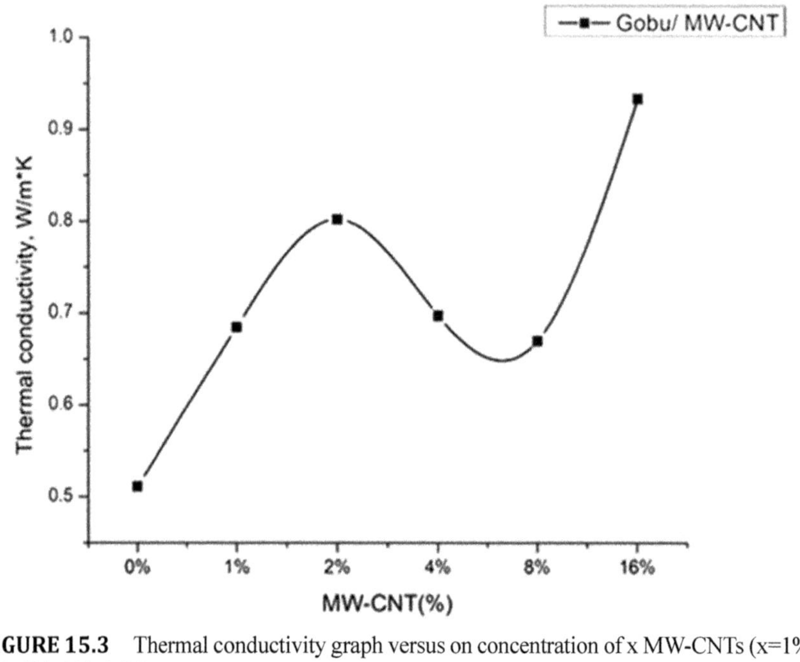

FIGURE 15.3 Thermal conductivity graph versus on concentration of x MW-CNTs (x=1%, 2%, 4%, 8%, 16%).

The thermal conductivity increases with increasing the concentration of MWCNTs. The results for 1%, 4%, and 8% concentration samples are almost identical. For the 2% sample, a high value was also observed (0.88 W/m*K) at 498 K temperature. The diagram form of TC is similar but depending on temperature, TC increases in some temperature ranges while decreasing in others. As expected with the addition of MWCNTs, the thermal conductivity of MWCNT/Gobu nanocomposites increased; however, the values of thermal conductivity are not very high.

15.4 CONCLUSION

The thermal conductivity of a horizontally aligned MWCNT/Gobu nanocomposite is $\kappa = 1.3$ W/(m*K) at 498 K for $x = 16\%$ of MW-CNTs. Despite high thermal conductivity of individual MWCNTs, inside natural bentonite matrices it significantly decreases, which can be explained mostly by phonon scattering from crossing MWCNT bundles, interfaces between MW-CNT and bentonite, bentonite structure and properties, bentonite porosity and network types of MW-CNTs formed inside composite.

KEYWORDS

- **MW-CNT**
- **nanocomposites**
- **thermal conductivity**
- **natural bentonite**

REFERENCES

1. Hendricks, T.; Caillat, T.; Mori, T. *Energies* **2022,** *15*, 7307.
2. Yung, K. C.; Lime, H. J. *J. Appl. Polym. Sci.* **2007,** *106*, 3587.
3. Han, Z.; Fina, A. *Prog. Polym. Sci.* **2011,** *36*, 914.
4. Yu, C.; Shi, L.; Yao, Z.; Li, D.; Majumdar, A. *Nano Lett.* **2005,** *5*, 1842.

5. Huseynov, A.; Mammadova, S.; Zeynalov, E.; et al. *Fullerens Nanotubes Carbon Nanostruct.* **2022,** *1*, 1.
6. Huseynov, A.; Israfilov, A.; Mammadova, S.; et al. *J. Compos. Mater.* **2019,** *2*, 1.
7. Liu, S.; Han, J.; Gao, Q.; et al. *Mater. Express* **2020,** *10*, 25864.
8. Kamal, A. M.; Basim, A. J. *Appl. Clay Sci.* **2007,** *37*, 90.
9. Sari, A.; Karaipekli, A. *Mater. Chem. Phys.* **2008,** *109*, 459.
10. Li, M.; Guo, Q. G.; Nutt, S. *Sol. Energy* **2017,** 146, 1.
11. Mammadova, S.; Baba, T.; Mori, T.; Huseynov, A.; Zeynalov, E. Thermal Transport Properties of MWCNT Based Natural Azerbaijani Bentonite Ceramic Composites. International Journal of Thermophysics. 2023, 44. 10.1007/s10765-023-03203-4.

CHAPTER 16

Preventing Extra Produced Water from an Oil Reservoir by Injecting Polymer Gel into this Reservoir

ABDOLLAH ESMAEILI, LAURA GARZÓN, RÔMULO SIMÕES ANGÉLICA, MARYAM DEHGHANI, CLAUDIO REGIS DOS SANTOS LUCAS, and AFONSO CESAR RODRIGUES NOGUEIRA

Universidade Federal do Pará (UFPA), Belem, Brazil

ABSTRACT

In this chapter, considering reservoir conditions (temperature and formation water, etc.), a polymer gel suitable for preventing extra produced water from an oil reservoir, based on the main characteristics of the polymer and the network-forming agent, has been researched. As polymer gels are prepared using two main materials, polymer and cross-linking agent, it is necessary to make base polymer and cross-linking agents in different ranges of temperature and characteristics of water. To be optimized, different samples of polymer were made and the properties of these polymers were studied in terms of functional groups, rheological properties, and stability of the gel network. After the initial investigations on the structure of the polymers, from the point of view of stability against the ions in the formation water and pH of the environment, the suitable polymer was identified and the gel obtained from bonding of the suitable polymer with the cross-linking agent chromium (III) acetate was investigated. Finally, the suitable composition for making the optimal polymer gel for water control operations was determined.

Advances in Engineering Materials: Preparation for Sustainable Process Development. Tamara Tatrishvili and A. K. Haghi (Eds.)

16.1 INTRODUCTION

In order to select the type and amount of functional groups attached to acrylamide copolymer, and the optimal combination of polymer and cross-linking agent, in this paper, experiments were conducted in two phases, which are as follows:

a) Examining the structure of the copolymer from the point of view of stability in the electrolyte environment and investigating its rheological properties in order to calculate the viscosity and molecular weight.
b) Examining the structure of the gel obtained from the designed polymers and the cross-linking agent, from the point of view of the setting time and the shrinkage time of the gel (in order to investigate the role of functional groups in the gel structure).

In order to localize the polymer gel system and considering the successful application of this technology, it is imperative to make a suitable copolymer and then design a gel suitable for the conditions of reservoirs. Therefore, it is desired to design and manufacture a suitable copolymer that can compete with other samples.

At this stage, tests were performed on the samples in order to investigate the viscosity and compare the molecular weight of the polymers based on the Mark-Huink law, on the samples of 1% by weight of the received polymers. In addition, the stability of the polymer in the electrolyte environment and in acidic and basic solutions was investigated.

16.2 STRUCTURE OF DESIGNED POLYMERS

Based on the library studies, acrylamide copolymer with sulfonate and carboxylate functional group was selected as the suitable polymer for gel preparation. For this purpose, 5 groups of polymers named S, TO, DO, HPAM, and PAM were made, and the members of each group have the same materials (sulfonate, amide, and carboxylate groups) with different percentages so that in addition to choosing the type of functional group, the amount and composition of its percentage was also calculated and examined. In order to reduce the independent variable in the evaluation and selection, four control samples were designed and made in the experiments.

16.2.1 S FAMILY POLYMERS

The structure of these polymers consists of two monomers, which are acrylamide and 2-acrylamido-2-methyl propane sodium sulfonate. Due to the fact that 2-acrylamido-2-methylpropane sodium sulfonate monomer has a negative charge and acrylamide monomer is neutral, therefore, the amount of functional group (with negative charge) determines the amount and type of polymer charge. The amount of functional groups in each of the S family copolymers and their molecular weight are different in order to investigate their role in the performance of the resulting gel.

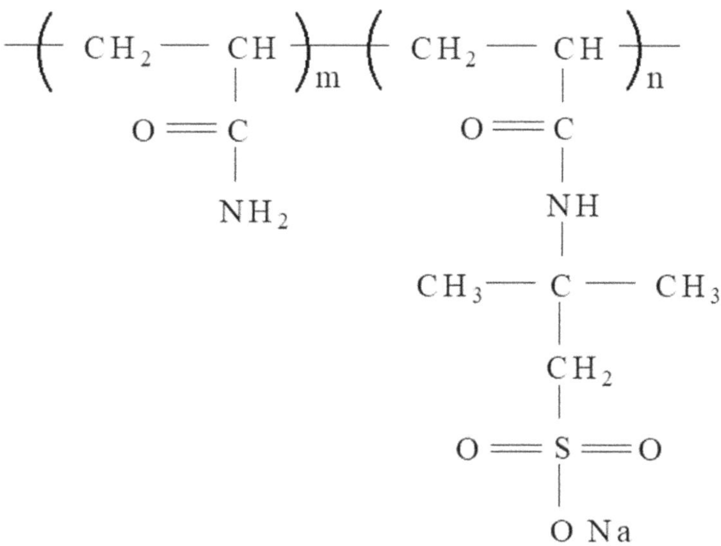

FIGURE 16.1 General structure of S family polymers.

16.2.2 TERPOLYMER TO1

This material is a terpolymer composed of acrylamide, sodium acrylate, and sodium 2-acrylamido-2-methylpropane sulfonate. The amount of acrylamide is 0.4 mol and sodium acrylate is 0.3 mol, and the amount of sodium 2-acrylamido-2-methylpropane sulfonate is 0.3 mol (30% sulfonic group). This polymer is also anionic due to the presence of a functional group with a negative charge and its hydrolysis degree is 10%.

16.2.3 DO FAMILY POLYMERS

The polymers of this family are composed of three monomers: acrylamide, sodium acrylate, and sodium 2-acrylamido-2-methylpropane sulfonate. This terpolymer is also anionic due to the type of functional groups.

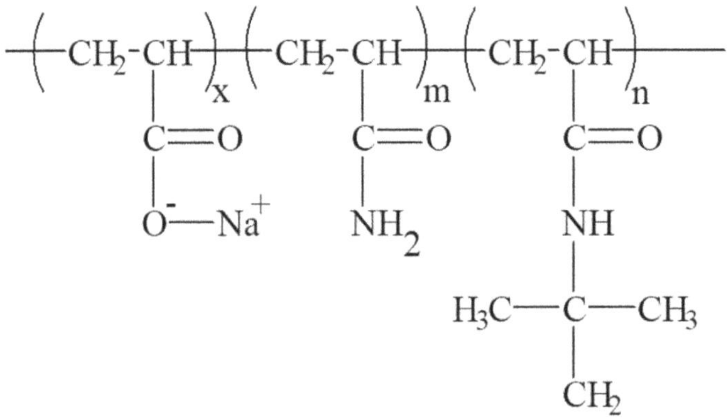

FIGURE 16.2 General structure of DO family polymers.

16.2.4 HYDROLIZED POLYACRYLAMIDE (HPAM) (FAMILY POLYMERS)

The polymers of this family are composed of two monomers, acrylamide and sodium acrylate. This copolymer is also anionic due to the carboxylate functional group.

16.2.5 POLYACRYLAMIDE PAM

In order to investigate the effect of polyacrylamide alone, a PAM sample was made and studied. A number of amide groups ($CONH_2$) of polyacrylamides are converted into carboxylate groups (COO-) at high temperature (90°C) and also in alkaline or acidic environment (such as sodium hydroxide). The fraction of amide groups that turns into carboxylate groups is called the degree of hydrolysis. In this case, this polymer is called hydrolyzed polyacrylamide. Therefore, due to the high temperature of the test environment (90°C) and the passage of time, there is a possibility of polymer hydrolysis.

16.3 RHEOLOGICAL TESTS ON DESIGNED POLYMERS

In order to investigate the properties of pure polymer solution, rheological tests were used. First, the initial viscosity of the polymer solution was calculated from the point of view of injectability and predicting the molecular weight of each polymer using Brookfield tests. For this purpose, 1% by weight solution of all polymer samples was subjected to constant stresses of 10, 30, 50, 100, and 120 1/S. According to the changes in the viscosity of the polymer according to the changes in the stress on the sample, and through the Mark-Houwink equation, the molecular weight of the polymer samples can be compared, and in addition, their penetration ability can be determined.

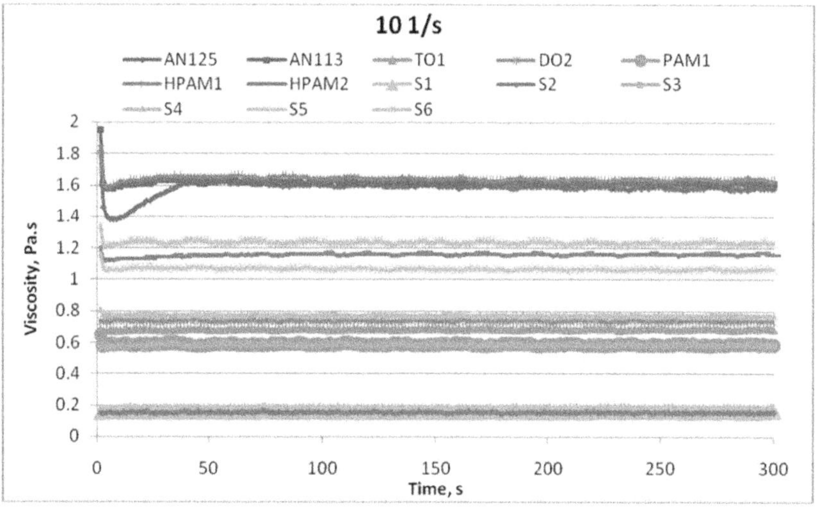

FIGURE 16.3 Viscosity changes of the samples in terms of time at a constant cutting rate of 10 1/s.

16.4 IMPROVEMENT OF GEL PROPERTIES

As mentioned, the most important polymer used in the oil industry to increase oil recovery, improve oil recovery, and block water is from the polyacrylamide family (with different degrees of hydrolysis and molecular weight). These polymers are usually cheap and can be converted into gel by organic and metal network-forming agents. The speed of converting these polymers into gel is strongly dependent on the density of the bonding agent. Other factors such as temperature, monovalent, divalent and trivalent ions, pH, polymer concentration, and cross-linking agent can also affect the speed and time of gel formation.

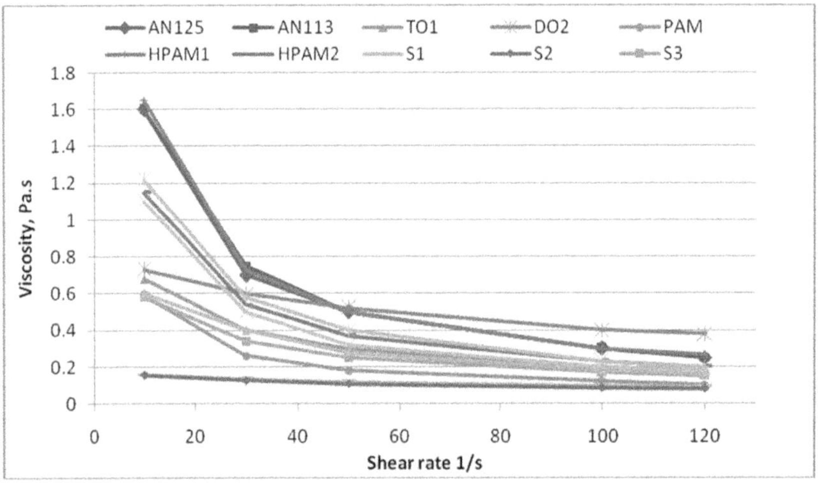

FIGURE 16.4 Comparison of changes in viscosity of samples according to time based on changes in cutting rate.

16.5 CONCLUSION

The main objective of the experimental work is to observe injected polymer gel effects on oil recovery for a Malaysian oil sandstone reservoir through polymer gel injection. The results from each part of this experiment can be compared with each other. Cores can be taken from the same well at approximately the same depth or different wells and different depths. For cores, the aging and flooding temperature must be at reservoir temperature. Consequently, polymer gel injection process will be presented as an efficient method for extra water production in this Malaysian oil reservoir.

ACKNOWLEDGMENT

The authors wish to express their sincere thanks and gratitude to Universidade Federal do Pará (UFPA) for their helps and encouragements in connection with this study. We are grateful for all their assistances.

KEYWORDS

- **polymer gel**
- **produced water**
- **oil reservoir**
- **cross-linking agent**
- **rheological properties**

REFERENCES

1. Bailey, B.; Crabtree, M.; Tyrie, J.; Elphick, J.; Kuchuk, F.; Romano, C.; Roodhart, L. Water Control. *Oilfield Rev.* Spring 2000.
2. Saxon, A. Keeping Water in its Place Middle East Well Evaluation Review No.19, 1997.
3. Sparlin, D. D.; Hagen, R. W. Jr. p "Controlling Water in Producing Operations, Part I: Where it Comes from and the Problems it Causes". World Oil, March 1984, pp 67–74.
4. Arnold, R.; et al. Managing Water- from Waste to Resource. Oil Field Review, Summer 2004.
5. Nasr-El-Din, H. A.; Taylor, K. C. Evaluation of Sodium Silicate/Urea Gels used for Water Shut-Off Treatments. *J. Petroleum Sci. Eng.* **2005,** *48*, 141–160.
6. Jurinak, J. J.; Summers, L. E. Laboratory Testing of Colloidal Silica Gel for Oilfield Application, Paper SPE 23581, Supplement to SPE 18505, Conoco Inc., 1991.
7. Herring, G. D.; Milloway, J. T. ARCO Alaska Inc.; Wilson, W. N.; ARCO Oil and Gas Co. In *Selective Gas Shut-Off using Sodium Silicate in the Prudhoe Bay field, AK.*, Paper SPE 12473 Presented at SPE Formation Damage Control Symposium, Bakersfield, California, Feb13–14, 1984.
8. Vinot, B.; Schechter, R. S.; Lake, L. W. Formation of Water Soluble Silicate Gels by the Hydrolysis of a Diester of Dicarboxylic Acid Solubilized as Microemulsions. *Soc. Pet. Eng. Reserv. Eng.* **1989,** *4* (3), 391–397.

CHAPTER 17

Extracting Optical Parameters of Biological Tissues by the Optical Spectroscopy of Skin In Vivo

VAKO MARCHILASHVILI[1], GIORGI KACHLISHVILI[1], ZAZA JALIASHVILI[1], TAMAZ MEDOIDZE[1], ALEXANDER GAMKRELIDZE[2], OLEG KHARSHILADZRE[2], LUKA TSULUKIDZE[2], ANA PETRIASHVILI[1], MARIAM CHUBINIDZE[3], IRAKLI KIRTADZE[3], ZAZA SOPROMADZE[3], ZAZA MELIKISHVILI[1], FEDOR ILKOV[4], and SUBRATA DUTTA[5]

[1]*Department of Optics and Spectroscopy, Institute of Cybernetics, Georgian Technical University, Tbilisi, Georgia*

[2]*Faculty of Exact and Natural Sciences, Tbilisi State University, Tbilisi, Georgia*

[3]*Jo Ann University Hospital, Tbilisi, Georgia*

[4]*IlkovF Consulting, Morgan Hill, CA, USA*

[5]*SKD Consulting, Glenn Dale, MD, USA*

ABSTRACT

We present results from in vivo skin reflectance spectroscopy with high spectral resolution. Subtle details of the reflected signal due to the biochemical and structural status of the skin (vascular network + bloodless skin) were detected and analyzed.

Optical models of the skin are analyzed and all the optical parameters that may be important in the study of human health are presented.

Advances in Engineering Materials: Preparation for Sustainable Process Development. Tamara Tatrishvili and A. K. Haghi (Eds.)

An original experimental setup has been designed and built for in clinical trials.

A spectroscopic study of healthy people with angiology problems has been performed and the effectiveness of in vivo experiments has been demonstrated.

A new parameter is proposed for angiologic practice—the optical ankle brachial index (O_{ABI}), which effectively describes the state of human health, as well as the change in this state during surgical intervention.

17.1 INTRODUCTION

17.1.1 SPECTROPHOTOMETRIC METHODS IN MEDICINE

Spectrophotometric methods, recognized for their high informativeness and precision,[1,2] serve as the cornerstone of laboratory analysis for biological specimens and form the foundation for various photometric techniques in clinical diagnostics that require sample collection. Clearly, it is far more convenient if we can conduct tests that do not involve invasive sampling procedures, thus avoiding discomfort or pain for the patient. This is the path modern optical-spectroscopic research in medicine should explore: the development of noninvasive approaches to study biological objects, all the while preserving the information richness and precision of in vitro methods.

Noninvasive optical techniques for assessing the condition of the human body are presently extensively employed in various fields of medicine, including cardiology, oncology, and gastroenterology, among others. These methods enable the monitoring of temporal changes in the body's state as well as structural and morphological alterations, all without perturbing the object under examination. You can find further details and examples in references.[3–8]

The process of assessing biological tissues' functional status involves two steps. First, optical parameters of the object are derived from the collected measurement data. Second, the inverse problem is addressed, aiming to recover the specific parameters of interest (such as microphysical or biophysical properties) from the determined optical parameters.

To achieve the required level of accuracy for practical applications, it becomes essential to employ radiation transfer theory methods that avoid relying on various simplifying approximations about the dominant interaction processes. However, it is important to note that these methods often come with substantial computational costs, which can make it impractical to interpret experimental data in real time. This limitation is particularly evident when utilizing Monte Carlo (MC) method as a tool widely used on the second step.[9]

Thus, under conditions of a priori uncertainty, the effective use of optical-physical measurement methods in noninvasive diagnostics of biological objects requires the development of methodological approach for interpretation of the measurements. Below we consider aspects of solving this problem in the diagnostics of biological objects.

17.1.2 DIFFUSE REFLECTANCE SPECTROSCOPY

Diffuse reflectance spectroscopy (DRS) is a noninvasive technique used to analyze tissue composition by measuring the amount of light reflected from the tissue. This provides insights into scattering and absorption within the tissue being examined. Understanding this technique requires familiarity with the optical properties relevant to the propagation of photons in tissue, and how these properties are connected to human health. The noninvasive properties of light, when applied for the examination and identification of irregularities in human tissues, present a highly appealing prospect for the advancement of novel diagnostic techniques. Among these techniques, DRS methods stand out as being relatively straightforward to deploy and, importantly, safe for the patient's well-being.[10–14]

In most well-established DRS methods, optical fibers serve as the conduit for delivering radiation to the target object and for collecting the backscattered light. In principle, these measurements can be conducted using commercially available spectrometric instruments, comprising a broadband radiation source connected to an optical fiber for delivering radiation to the object and a multichannel spectrometer linked to optical fibers to capture the scattered radiation from various distances around the irradiation area. However, alternative configurations of measurement devices are also feasible.[15,16]

The general setup of a device for spectral measurements of backscattering signals, including those with spatial resolution, is illustrated in Figure 17.1.

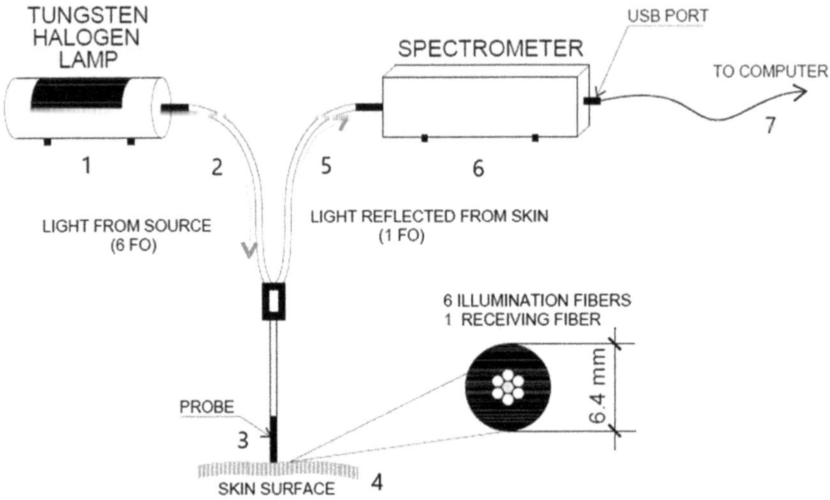

FIGURE 17.1 Block diagram of a device for spectral measurements of backscattered light. 1—light source, 2—light-delivering fibers, 3—fiber optic probe, 4—object under study, 5—light-receiving fibers, 6—CCD or CMOS spectrometer, 7—block for controlling the light source and processing measurement information.

The light source (labeled as 1 in the figure) can utilize a broadband emitter, such as a white light LED or a tungsten-halogen lamp, or it can consist of a collection of narrowband emitters, including laser or LED sources. Light generated by the radiation source 1 is directed into the fiber optic probe 3, which guides it to the target object under investigation (labeled as 4). The radiation scattered by skin is collected by the light-receiving fibers 2, which transport it to the spectrometer (labeled as 6). The signals from the spectrometer are processed by a computer (labeled as 7), on which a specialized program carries out a quantitative analysis for determining the parameters of the object.

The detected signals represent a combination of tissue scattering and absorption properties, and contain information on the tissue's structural and morphological parameters, as well as the concentrations of optically active chromophores (such as oxyhemoglobin, deoxyhemoglobin, melanin, bilirubin, etc.), along with details about the concentrations and size of effective tissue scatterers.

Overall, DRS's noninvasive nature and its ability to provide quantitative information about tissue properties make it a potentially valuable tool for diagnosing and monitoring various health conditions. However, more research is needed to establish its clinical utility.

The complexity of biological tissue requires creation of specialized models that take into account the composite structure of the tissue and multistep photon interaction process.

In the article, there is a strong emphasis on the importance of understanding biological tissue, particularly in terms of its structural complexity. To gain a comprehensive understanding, it is crucial to break down this complexity into manageable parts and study the interactions between these components.

The article uses models that are derived from the Feynman interpretation of quantum mechanics. Specifically, it mentions analytical methods such as the path integral model and the diffusion approximation model, which are used to represent these interactions.[17–19] Additionally, the article also highlights the utilization of Monte Carlo simulations[9] as a valuable tool in understanding light–tissue interactions. These models and simulations help researchers gain insights into the complex behavior of biological tissue under exposition by optical radiation.

The structure and purpose of the work are defined as follows:

1. **Physical Basis:** a brief overview of the physical methods.
2. **Optical Models:** Optical models for skin are discussed. Most appropriate model for our application is suggested.
3. **Data Collection:** Details on collecting skin spectra from healthy individuals, processing the data, and deriving optical parameters of the skin.
4. **Clinical Experiments:** Experiments in a clinical setting to validate the method's effectiveness are described.

The ultimate goal of this work is to create a noninvasive, efficient, and affordable device for monitoring human health.

17.2 PHYSICAL FOUNDATION OF METHOD

17.2.1 FROM PHYSICS TO MEDICINE

The interaction of light with biological tissue is complex and includes the optical processes of elastic scattering, absorption, and emission.[20]

The wavelength of optical radiation also plays an important role in interactions with biological tissues for the following reasons[21]:

1. **The Water Optical Transparency Window:** Tissues primarily consist of water, which strongly absorbs light in the ultraviolet and longer infrared ranges. Melanin also absorbs shorter wavelengths of light. However, there exists a spectral "window" in water's absorption spectrum that allows visible and near-infrared light (600–900 nm) to penetrate more effectively. This makes visible or near-infrared wavelengths preferred choices for optical spectroscopy of biological tissues.[22]
2. **Isobestic Wavelength:** Oxyhemoglobin (HbO_2) and reduced hemoglobin (Hb) exhibit significant differences in absorption, except at specific isobestic wavelengths. Wavelengths where the absorbency of HbO_2 is equal (isobestic points) are 548, 568, 587, and 805 nm.[23] When measurements are conducted at an isosbestic wavelength the signal remains largely unaffected by changes in blood oxygen saturation.
3. **Tissue Penetration Depth:** The depth to which light can penetrate tissue, for a given intensity of optical radiation, depends on the wavelength used. A common metrics for the penetration depth is attenuation of light by "e" times. For example, intensity attenuates by e times at 460 nm at 0.5 mm and at 640 at 1.5 mm.[24] In optical spectroscopy, the sample volume is also determined by probe design, and can extend to approximately 1 cm^3 for transmission-mode systems. This makes PPG a powerful tool for assessing capillary nutritional blood flow and thermoregulatory blood flow through arteriovenous anastomosis shunt vessels.

To characterize the specificity of the photon path in biotissues, let us consider monochromatic radiation propagating in biological tissue. Such a light will be attenuated due to multiple elastic scattering and absorption.[20] The elastic scattering is a process that changes only direction of the incident radiation, so in this approximation, we can use the Lambert-Beer law[20] and write down the changes in the radiation intensity along a given trajectory ℓ

$$I(\ell, \lambda) = I(0, \lambda) \exp\left(-\int^{\ell} \mu \, d\ell'\right),$$

or

$$R(\ell, \lambda) \equiv I(\ell, \lambda) / I(0, \lambda) = \exp\left(-\int^{\ell} \mu \, d\ell'\right) \tag{17.1}$$

In this expression $R(\ell, \lambda)$ is the diffuse reflection coefficient, $I(\ell, \lambda)$ is the intensity of the beam after traveling the ℓ distance in the biological tissue with attenuation coefficient $\mu = \mu_a + \mu_s$ (μ_a – absorption; μ_s—scattering) and $I(0, \lambda)$ is the intensity of the beam incident on the surface of the tissue. For a homogeneous biological tissue from (17.1) we obtain

$$R(\ell, \lambda) = \exp(-\mu \ell) \tag{17.2}$$

Now, let us start with estimating the value of ℓ in nonabsorbing medium, $\mu_a << \mu_s$, when approximation $\mu_a = 0$ is a good one. Then, the light distribution could be described by the solution to the radiative transfer equation (RTE), and propagation of radiation can be described by a model of diffuse propagation of light in a homogeneous medium.[17–19] Results of optical radiation propagation modeling from the source to the detector in a semi-infinite medium are similar to shown on Figure 17.2. The banana-shaped region reaches the deepest part of the sample at its middle.[25] Such a form is derived from the analytical diffusion theory (DT) solution of the RTE.[26,27]

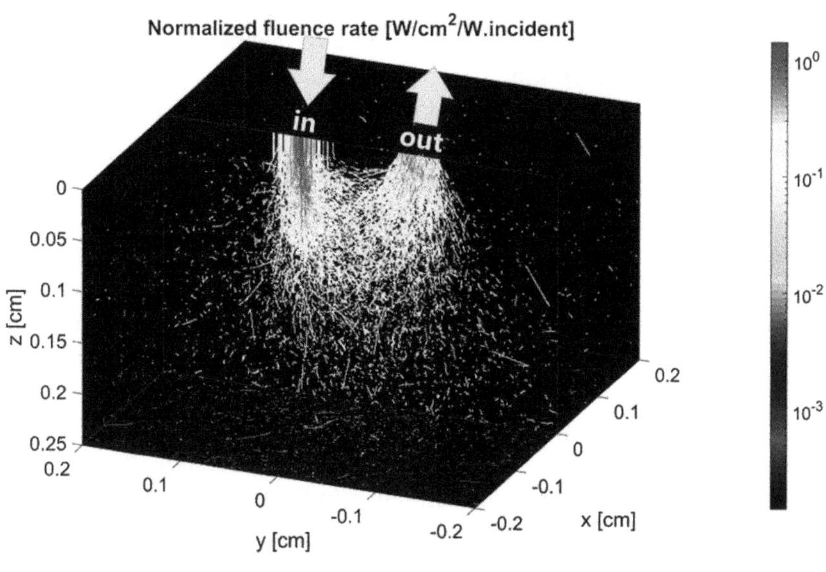

FIGURE 17.2 Monte Carlo picture of optical radiation migration in tissue performed by MCmatlab described in Ref. [28].

Source: Adapted from Ref. [28]

The result of solving the radiative transfer equation (RTE), with Monte Carlo (MC) methods[28] is shown in Figure 17.2. In these methods the light is simulated as photon packets that are gradually absorbed as they travel through the medium and will randomly undergo scattering at a rate dependent on the local optical properties of the medium.

To summarize, the above methods of modeling confirm intuitively understood banana-shape propagation of the optical radiation propagation from source to detector—through the bio-tissue. The depth of the banana-shaped region depends on the radiation wavelength and tissue properties. Thus, such a method of optical tracking technique, which in fact is an optical tomography, is a good tool for examination and visualization of the internal structures of biological tissue.

17.2.2 OPTICAL MODEL OF THE SKIN

In contemporary scientific literature, there is a substantial body of work dedicated to modeling the behavior of light as it travels through biological tissues and computing their characteristics related to diffuse reflection (DR). In what follows, we will consider model described in Ref. [29] which, in our opinion, has all the main features that determine the process of light interaction with skin.

For modeling purposes, the skin is simplified into a horizontally homogeneous medium with three layers: the stratum corneum, epidermis, dermis. The later includes the papillary and reticular dermis, as well as surface and deep vascular plexuses.[30–34] Since the stratum corneum has minimal optical thickness, it has a negligible contribution into the DR. Consequently, it is typically grouped with the epidermis in modeling. Hence, the mathematical model utilized for skin in this study simplifies its structure into two flat layers, each significantly differing in composition and optical properties. The upper layer corresponds to the epidermis, with a thickness L_{epi} typically ranging from 50 to 130 μm, while the lower layer represents the dermis, often regarded as optically infinitely thick. It is worth noting that the refractive indices of both skin layers are assumed to be identical, denoted as n_{skin}. Consequently, Fresnel reflection of radiation exclusively takes place at the interface between the skin and the surrounding medium. Furthermore, we assume the same value of n_{skin} for the spectral range of interest taking into account negligible change of its variation contribution compared to the uncertainties associated with its absolute value. Additionally, we consider the tissue under the skin to be minimally absorbing.

With the above assumptions, the relationship between its absorption coefficient μ_t [cm^{-1}] and the wavelength λ [nm] could be estimated using the following expression:

$$\mu_t(\lambda) = 0.244 + 85.3 \exp[-(\lambda - 154)/66.2] \quad (17.3)$$

The "effective" scatterers within the tissue predominantly consist of collagen and elastin fibers in the dermis (arranged in bundles), and keratinocytes and melanocytes in the epidermis. For calculating the scattering phase function of those scatterers, we will employ the Henyey–Greenstein function. The spectral variation of the anisotropy factor (average cosine) of this function can be characterized by the empirical relationship:

$$g(\lambda) = 0.7645 + 0.2355\,[1 - \exp(-(\lambda - 500)/729.1)] \quad (17.4)$$

The spectrum of the transport scattering coefficient $\mu_s(\lambda)$ for biological tissues in the visible and near-infrared ranges is determined by combining the spectra $\mu_s(\lambda)$ of Mie and Rayleigh scatterers with sizes $d \geq \lambda$ and $d < \lambda$, respectively,[31]

$$\mu_s(\lambda) = \mu_s(\lambda_0)\,[\,\rho_{Mie}(\lambda_0/\lambda)^{x} + (1 - \rho_{Mie})(\lambda_0/\lambda)^{4}] \quad (17.5)$$

where λ_0 is set as 400 nm, ρ_{Mie} represents the fraction of Mie scattering, and x serves as the parameter governing the spectral dependence of the Mie transport scattering coefficient. The value of x is dependent on the size of the scatterers and their refractive index.

The volume absorption coefficients of the skin layers can be computed by applying established principles that involve summing optical quantities. Specifically, it is determined as the sum of the absorption coefficients of individual components, with the weights assigned according to their respective volume concentrations. In the case of the epidermis, the spectrum of the absorption coefficient can be calculated based on the volume concentrations of melanin f_{mel} and water w_{epi} present within it:

$$\mu_{epi}(\lambda) = f_{mel}\,\mu_{mel}(\lambda) + w_{epi}\,\mu_w(\lambda) + (1 - f_{mel})\,\mu_t(\lambda) \quad (17.6)$$

where μ_{mel} and μ_w are the absorption coefficients of melanin and water.

The absorption coefficient of the dermis is influenced by the presence and concentrations of capillaries with blood (f_{blood}), water (w_{derm}), and bilirubin (C_{bil}),[31,32]

$$\mu_{\text{derm}}(\lambda) = f_{\text{blood}}\, \alpha\, k_{\text{blood}}(\lambda) + w_{\text{derm}}\, k_{\text{wat}}(\lambda) + (1 - f_{\text{blood}}) [\mu_t(\lambda) + \ln(10)\, (C_{\text{bil}} / \mu_{\text{bil}})\, \varepsilon_{\text{bil}}(\lambda)], \tag{17.7}$$

where ε_{bil} is the molar absorption coefficient of bilirubin (in $\text{cm}^{-1}/(\text{mol/L})$), $\mu_{\text{bil}} = 585$ g/mol is the molar mass of bilirubin, μ_{blood} is the absorption coefficient of blood, and α is the correction factor that takes into account the effect of localized absorption of light by blood vessels.[33]

For randomly arranged capillaries with a diameter D, the coefficient α can be found from the following expression:

$$\alpha = (4/3)^{1/2} \{1 - \exp[-\pi\, \mu_{\text{blood}}\, D\, (1 - 0.43\, \mu_{\text{blood}}\, D) / (4/3)^{1/2}] / \pi\, \mu_{\text{blood}}\, D\} \tag{17.8}$$

The absorption coefficient of blood, $\mu_{a\ \text{blood}}$, is expressed as a sum of the absorption coefficients associated with various components, including hemoglobin (both oxidized and unoxidized forms) and bilirubin, each of which is wavelength dependent:

$$\mu_{\text{blood}}(\lambda) = \ln(10)\, (C_{\text{t Hb}} / \mu_{\text{Hb}})\, [S\, \varepsilon_{\text{Hb}}(\lambda) + (1 - S)\, \varepsilon_{\text{HbO2}}(\lambda)] + \ln(10)\, (t_{\text{dif}}\, C_{\text{bil}} / \mu_{\text{bil}})\, \varepsilon(\lambda) \tag{17.9}$$

Here:

- C_{tHb} [g/L] represents the concentration of total hemoglobin in the blood.
- $\mu_{\text{Hb}} = 64{,}500$ g/mol is the molar mass of hemoglobin.
- $\varepsilon_{\text{HbO2}}$ and ε_{Hb} are the molar absorption coefficients of oxyhemoglobin and deoxyhemoglobin, respectively, and are typically measured in units of $\text{cm}^{-1}/(\text{mol/L})$.
- S represents the saturation level of blood with oxygen, which is the fraction of oxidized hemoglobin in the total hemoglobin.
- $t_{\text{dif}} = 5$ is a concentration ratio, specifically the ratio of bilirubin concentration in the blood to that in the surrounding tissue.

To complete the optical model of skin, we need to establish the range of values for each of the following parameters:

1. Refractive index of whole skin (n_{skin}).
2. Spectral dependence parameters of the transport scattering coefficient, such as $\mu_s(\lambda_0)$, ρ_{Mie}, and x.
3. Thickness of the epidermis (L_{epi}).
4. Volume concentrations of melanin and water in the epidermis (f_{mel} and w_{epi}, respectively).

5. Volume concentrations of capillaries with blood, water, and bilirubin in the dermis (f_{blood}, w_{derm}, and C_{bil}, respectively).
6. Average diameter of capillaries (D).
7. Concentration of total hemoglobin in blood (C_{tHb}).
8. Saturation level of blood with oxygen (S).

We also assume that those parameters vary independently of each other.

17.2.3 PHOTON MIGRATION MODEL WITH EXPLICIT SOURCE-DETECTOR SEPARATION DEPENDENCE

Below we consider the models suitable for specifics of DR measurement with a fiber optic probe.[34] Previous calculations of DR from skin show that the range 0–30° in angles of incidence θ of radiation onto skin has no effect on the value of the DR.[35] Thus, with the use of fibers with NA = 0.3 as probes, the results of measurements of the DR values are determined only by distances between probes and skin's parameters.

17.2.3.1 SINGLE-LAYER MODEL

In accordance with the principle of optical "adding" of the layers,[16] a simplified model of biological tissue, particularly skin, can be considered a single layer in a multi-layer environment. In some frameworks, especially when dealing with basic simulations or simplifications, considering biological tissue like skin as a single layer has its merit by reducing the complexity of the model. However, real biological tissue, like skin, is a complex, multi-layered structure with various functions, and the simplification limits applicability of such model conclusions.

17.2.3.2 MULTI-LAYER MODEL

As was pointed out above, skin is composed of several distinct layers, each with its unique properties and functions. The main layers of the skin are the epidermis, dermis, and subcutaneous tissue (hypodermis). Each layer serves specific functions: (1) **Epidermis:** The outermost layer provides a barrier against environmental factors, contains melanocytes that produce melanin (skin pigment), and houses various types of skin cells. (2) **Dermis:** The middle layer contains blood vessels, hair follicles, sweat glands, and connective tissue. It is responsible for providing strength and support to the skin. (3) **Subcutaneous tissue (hypodermis):** The innermost layer consists of fat and connective tissue, providing insulation and cushioning.

When working with a multi-layer model, it is possible to simulate more realistic behaviors of skin, such as heat transfer, moisture retention, and mechanical properties. This can be important in fields like dermatology, biomechanics, and medical device development.

The choice between a single-layer or multi-layer model depends on the specific goals of the research or simulation. In many cases, researchers and engineers use multi-layer models to capture the complexity of biological tissue more accurately. However, single-layer models can be useful for quick, simplified analyses in certain scenarios.

In the modeling of light propagation in biological tissue, the internal reflection parameter (κ) determines how light refracted and reflected as it enters the skin layers:

$$(1 + r_{id}) / (1 - r_{id}) \tag{17.10}$$

where r_{id}, and as shown in Ref. [36], is described with a high accuracy by a polynomial

$$r_{id} = -\,1.4399\, n_{rel}^{-2} + 0.7099\, n_{rel}^{-1} + 0.6681 + 0.0636\, n_{rel} \tag{17.11}$$

In the simplest case of diffusion approximation, one can consider the skin as a semi-infinite space, with $n_{rel} = n_{tis} / n_{ext}$, where n_{ext} and n_{tis} are the refractive indices of the air and medium, respectively. From the theory of radiative transfer, the analytical expression for the spatially resolved stationary diffuse reflection that defines the dependence of the coefficient of light diffuse reflection on distance ρ between points of radiation input into the medium and detection of the light emerging from the medium has the form[37]

$$\begin{aligned} R(\rho) = (a'/4\pi)\,[1/(\mu_a + \mu'_s) + (\mu_{eff} + 1/r_1)\exp(-\mu_{eff} r_1)/r_1^2 + \\ (1/(\mu_a + \mu'_s) + 2z_b)(\mu_{eff} + 1/r_2)\exp(-\mu_{eff} r_2)/r_2^2], \end{aligned} \tag{17.12}$$

where

$$\mu_{eff} = [3\mu_a(\mu_a + \mu_s(1-g))]^{1/2} = [3\mu_a(\mu_a + \mu'_s)]^{1/2},$$
$$r_1 = (z_0^2 + \rho^2)^{1/2}, \qquad r_2 = [(z_0 + z_b)^2 + \rho^2]^{1/2},$$
$$a' = \mu'_s / (\mu_a + \mu'_s), \qquad z_0 = 1/(\mu_a + \mu'_s).$$

The extrapolated boundary distance is given by

$$z_b = 2\,\kappa\, D_{dif},$$

where D_{dif} is the diffusion constant

$$D_{dif} = 1 / 3\,(\mu_a + \mu'_s).$$

All the parameters in expression (13) have spectral dependences.

17.3 EXPERIMENTAL METHODS AND MATERIALS

17.3.1 INSTRUMENTATION

In the initial experiment, we opted for the dorsal side of the distal phalanx on the ring finger and big toe of the left foot as measurement sites for a healthy volunteer. Subsequently, these same sites were selected for both pre- and postoperation subjects with angiological issues in our further experiments (Figure 17.3). The device used in the clinic is shown in Figure 17.4.

The diffuse reflectance spectra of skin were measured by means of a fiber probe. As shown in the insert of Figure 17.1, the fiber probe had six illuminating and one detection 200-micron diameter fibers bundled together. 6-fiber leg was connected to light source, and reflected light was fed into spectrometer via single-fiber leg (Figure 17.1).

The numerical aperture of the fibers is = 0.22 ± 0.02, which is equivalent to an acceptance angle of 24.8° in air.

We used Ocean Optics HL-2000 Tungsten Halogen Lamp as a broad-band incandescent light source with a color temperature of 3000 K. A miniature CCD Spectrometer, Qmini-VIS, from RGB Photonics was used to collect spectra. The native spectral range of the spectrometer is 400 nm to 760 nm. The entrance slit of the spectrometer is imaged on a 2500-pixel CCD array by F = 50 mm optics with no magnification. We have widened the entrance-slit width to 200 μm in order to increase the light collection efficiency of the spectrometer. That slit width in combination with a 600 l/mm grating and 50 mm focal length optics resulted in a spectral resolution of 5 nm. That resolution was adequate for our needs for two reasons. First, skin tissue chromophores absorption spectra change on scale of tens of nm, and we are not missing any relevant details for our measurements. Second, we had to bin pixels on the detector in order to increase the acquisition speed, and increase SNR. The binning matched the reduced effective resolution of spectrometers even with 200 um entrance slit.

All spectra were calibrated using a diffuse reflectance standard WS-1 from Ocean Optics; calibration spectra on the standard were routinely measured prior to data acquisition.

The wavelength calibration of the spectrometers was performed using an Hg (Ar) calibration lamp (HG-1 Mercury Argon Calibration Source, Ocean Optics). The signal-to-noise ratio (SNR), dependent on wavelength, and for the shortest exposure time (1 ms) used in experiment was not worse than 150:1.

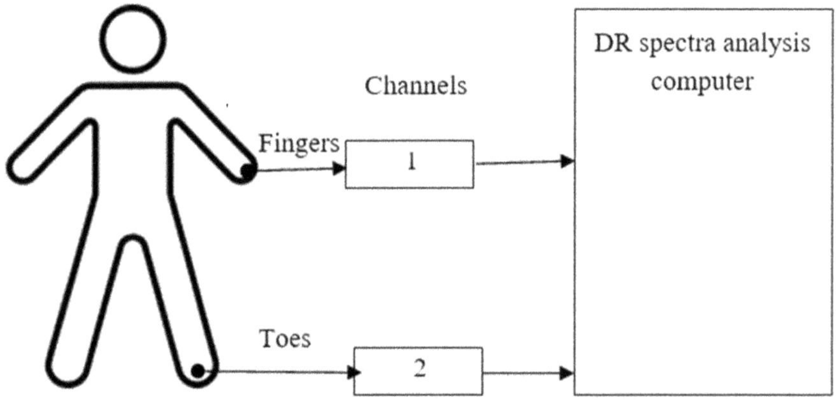

FIGURE 17.3 An overview of two-channel DR spectrometer and analysis system, giving capability for spectra to be compared between hand to foot sites.

The fitting was performed using the nonlinear least-squares minimization method, offered by the SciPy Python library. During the fitting process, the boundaries, depending on the meaning of the parameters, were either set to values that were somewhat outside their realistic magnitudes or given the widest logically possible ranges. For example, concentration variables were allowed to vary within the 0–1 range. The results displayed reasonable parameter values.

Before collecting data, we obtained a reference spectrum from a diffuse reflectance standard. Subsequently, all skin spectra were normalized to this reference spectrum from the illumination source.

To ensure adequate light coupling and enhance measurement repeatability, we gently pressed the fiber-optic probe against the skin surface during spectrum acquisition. Recognizing that probe pressure can impact the spectra, we maintained a consistent and minimal pressure on the skin throughout the acquisition process. We verified the measurement repeatability by acquiring spectra multiple times.

The influence of contact pressure on the diffuse reflectance of human skin has been thoroughly investigated and discussed in other papers.[38,39] We followed the recommendations from those studies regarding probe-skin pressure contact to ensure reliable measurements. Additionally, we mitigated any potential motion artifacts by carefully positioning the probe on the subject's skin.

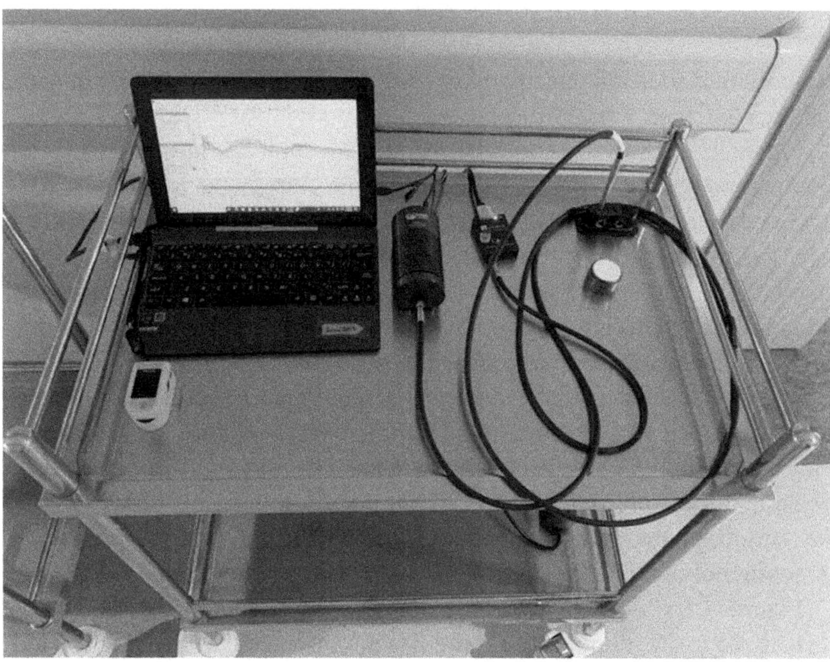

FIGURE 17.4 Clinical set-up.

It should be noted that optical fiber used for light transmission is flexible. During normal use, the bending angle change could result in the light intensity change. Normalizing the spectral curves to the maximum amplitude for each measurement, we mitigate possible variations in the ligth intensities. Wavelength dependence of the intensity losses is negligible. The parameter that determines wavelength-dependent loss is equal to $(0.16\ \lambda/R)^{2/3}$ (R is the bending radius).[40] In our case λ is about 10^{-7} m, R is about 10^{-1} m, and the dependence on wavelength can be neglected.

Another possible mechanism responsible for intensity loss is microscopic bend. Microbending in multimode fibers leads to the transfer of part of the energy from one mode to another. Microbending losses in such fibers do not depend on the wavelength.[40] Thus, the corresponding spectra, normalized to maximum amplitudes, represent an objective characteristic of the reflective properties of biological tissue.

17.4 RESULTS AND DISCUSSION

We have measured DR spectra taken on a ring finger of 41 healthy subjects, and on a toe from 9 healthy subjects.

Figure 17.5 shows average by 41 healthy subjects, normalized to the illumination source, reflection spectrum from ring finger taken with 1 ms exposure time with averaging over 100 exposures. For future comparison with spectra taken from other sites, this reflectance spectrum is normalized to the maximum amplitude value, $C_{nor} = 14.72144$. The distinguished blood (HbO_2 + Hb) absorption features, around 420 nm and between 515 nm and 615 nm, are clearly seen in the spectrum.

The DR spectra exhibit features characteristic to hemoglobin absorption around 420 nm, 540 nm, and 580 nm. Overall decrease in reflectance with increasing wavelength in this wavelength range is characteristic to melanin absorption. In the 625–750 nm wavelength range, absorption by skin chromophores is minimal and scattering effects dominate, resulting in approximately linear decrease of DR value with wavelength.

Figure 17.6 shows average by 9 healthy subjects, DR spectrum normalized to the illumination source, taken from toe n with 1 ms exposure time averaged over 100 exposures. The spetrum was normalized to its maximum amplitude ($C_{nor} = 13.39599$). Comparing the spectral shape in Figure 17.5 (finger) with the spectrum in Figure 17.6 (toe) reveals close similarity.

To compare the above experimental spectra with model prediction, we run, described in Section 17.2.1, MC procedure to model the DR spectra in vivo of human skin on the finger. Figure 17.7 shows the typical fit of the model to the measured skin DR spectra The modeling results are very close to the measured data. That validates the used MC model.

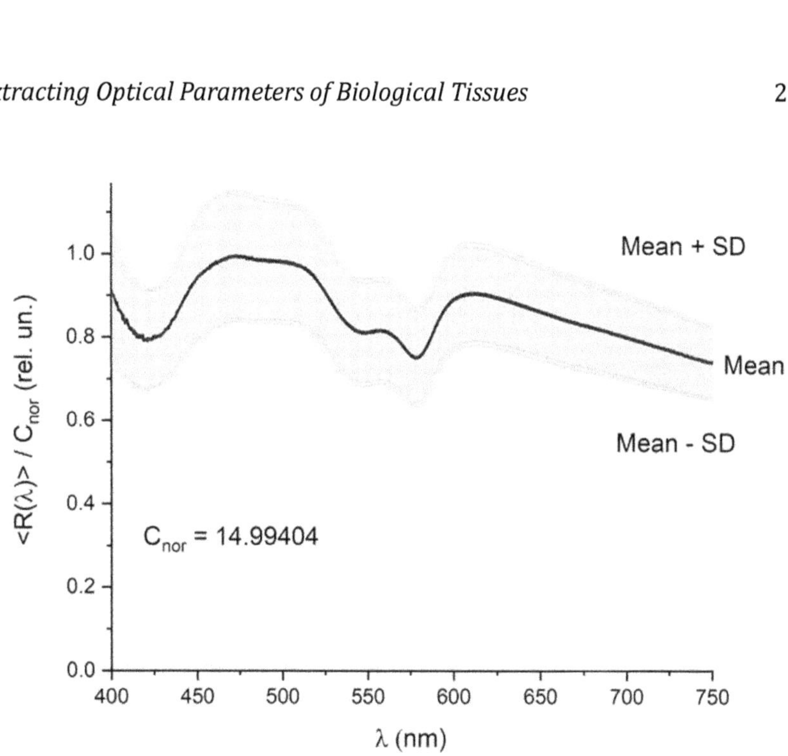

FIGURE 17.5 DR spectrum of the left-hand ring finger (middle line) averaged over 41 healthy subjects. The shaded area is marking ±1 SD.

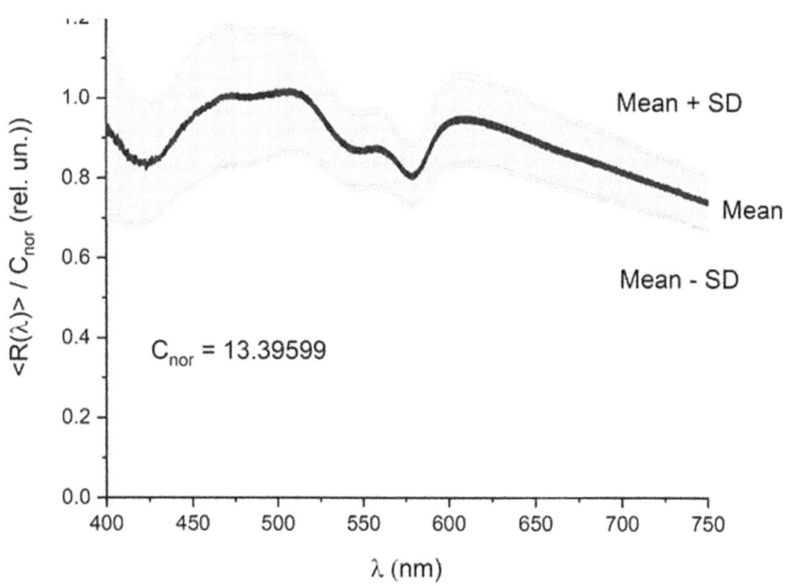

FIGURE 17.6 Averaged by 9 healthy subjects reflectance spectrum of toe (middle line) with the shaded area marking ±1 SD.

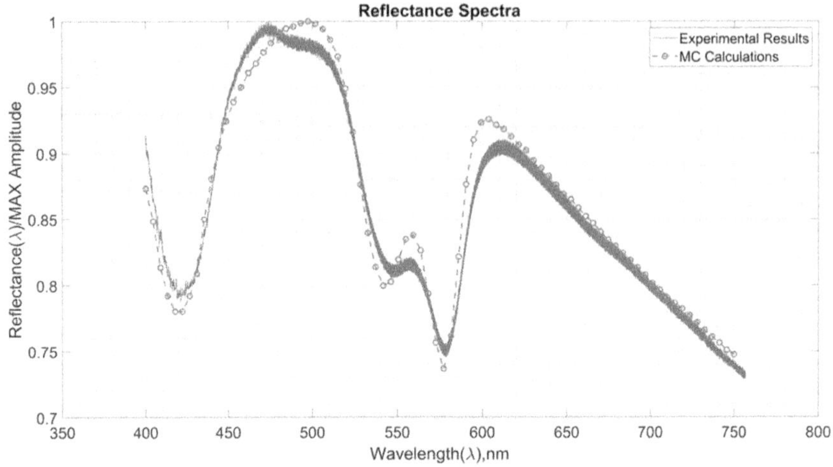

FIGURE 17.7 Typical diffuse reflectance spectra of healthy human skin measured in vivo. Blue line represents experimental data while red dotted line corresponds to MCmatlab model fit.

Table 17.1 summarizes the values of essencial skin parameter extracted from the measured on all volunteers. These values are in good general agreement with the results of previous studies reporting on the optical properties of human skin.[16]

TABLE 17.1 Comparision of Essential Skin Parameter Ranges, Previously Reported in Ref. [16], with those Obtained by us, using the Model Fitting to our Measurements on Healthy Subjects' Skin.

Variation ranges of parameters [16]		Obtained parameters by fitting	
Parameter	**Range**	**Epidermis**	**Dermis**
n_{skin}	1.4–1.5	1.45	1.40
$\mu_s(\lambda_0)$	3–11 mm^{-1}	6.67 mm^{-1}	4.36 mm^{-1}
x	0.5–1.0	0.689	0.562
ρ_{Mie}	0.1–0.6	0.51	0.50
f_{mel}	0.5–25%	1.3%	-
L_{epi}	50–150 μm	90 μm	-
f_{blood}	0.2–7%	-	5.5 %
S	40–98%	-	99%
C_{bil}	5–250 mg/L	-	140 mg/L
W_{epi}	20–40%	33%	-
W_{derm}	60–77 %	-	75 %

The diffrence between measured spectral shapes of DR taken on finger and on toe in healthy subjects is small (Figures 17.5 and 17.6). Hence, we propose to introduce an optical index, O_{ABI}, that is equvalent to ankle brachial index (ABI) which is widely used in clinical studies.[41] The ABI is calculated from the results of the blood pressure measurements in the upper and lower limbs. Health care providers calculate ABI by dividing the blood pressure in an artery of the ankle by the blood pressure in an artery of the arm. The ABI index is a rapid, noninvasive initial test that assesses the adequacy of circulation to the extremities. It gives clinicians information about whether the blood vessels in the extremities are damaged or narrowed, or whether peripheral artery disease has developed.[43] Using the ABI index, clinicians get an idea of how much blood flows into the lower extremities. In healthy individuals, the ABI index ranges from 0.9 to 1.2. An ABI index below 0.9 indicates an arterial disease, predominantly an obstructive process of the aorta or vessels of the lower extremities, the cause of which in most cases is atherosclerotic lesions. A high ABI index > 1.2 is also an indicator of pathological changes and primarily indicates calcification of the arterial wall and a decrease in its elasticity.

The value of the new $\mathrm{O_{ABI}}$ index, we are proposing, is presented in Figure 17.8 for a healthy person. As expected, the value of $\mathrm{O_{ABI}}$ varies very little with wavelentgh for a healthy person.

Similarly to the standard ABI index, we compared new $\mathrm{O_{ABI}}$ index values, before and after surgery. Figure 9 show $\mathrm{O_{ABI}}$ index values from a patient before and after surgery. A striking difference in the $\mathrm{O_{ABI}}$ index's values clearly indicates that the surgery was successful.

17.5 CONCLUSION

We have shown that, based on well-understood physics of the diffuse reflectance process, and availability of sophisticated models describing the process, it is possible to develop a straightforward methodology for evaluating the diffuse reflectance in biological tissues in a clinical setting and use the results for evaluating human health status.

As far as we know, this is the simplest and one of the most cost-effective methods for noninvasive evaluation of human health status, at least for the case described in this study.

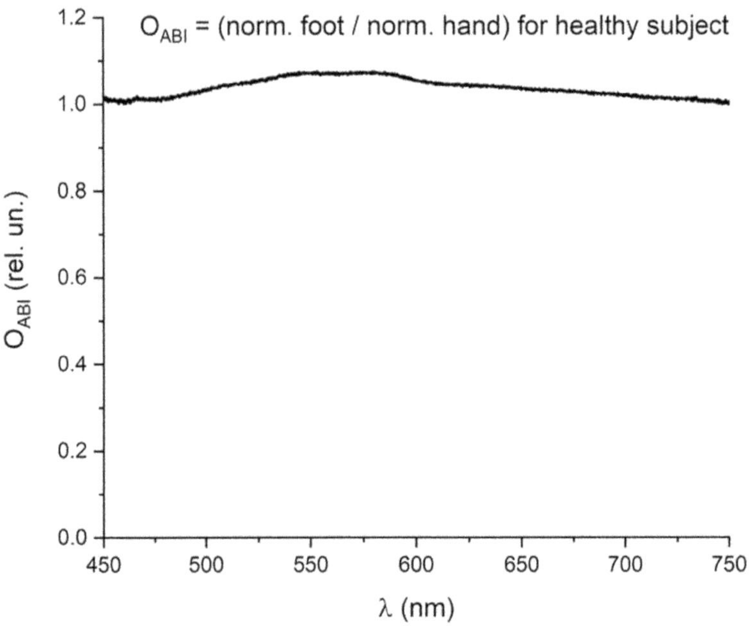

FIGURE 17.8 O_{ABI} for a healthy subject obtained by dividing mean value of DR spectra obtained on toe to the mean value of DR spectra from finger.

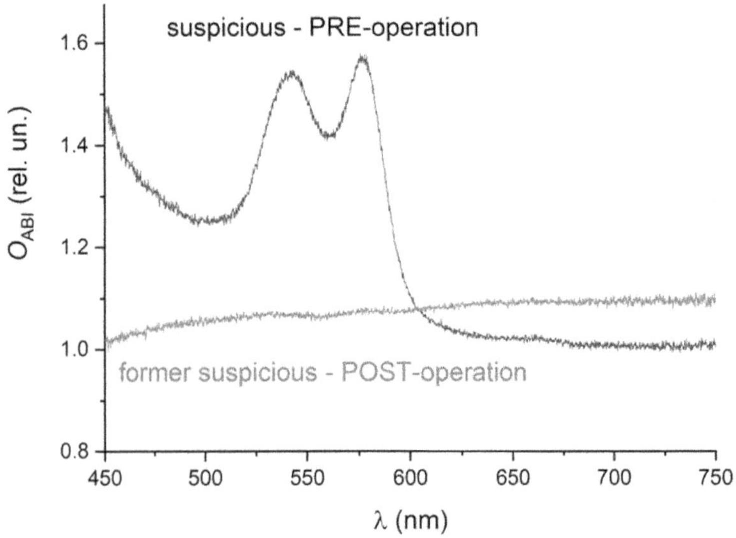

FIGURE 17.9 O_{ABI} values for patient before surgery in black trace, and after surgery in red trace.

The method of analyzing the DR spectra of human skin in vivo, described in this study, can be used in real time during a routine patients' screening, as well as during surgeries.

The described methods can also be used to determine in real time the parameters of biological tissue without contact of the measuring instrument with the patient's body.

In modeling diffuse reflectance, we have assumed tissue to be a two-layer medium: homogeneous epidermis layer and homogeneous semi-infinite dermis. Even though this is clearly a first approximation and may even appear as a gross oversimplification for biological tissue, its applicability and utility has been tested and demonstrated by a number of studies (see Ref. [16] and references cited in there). In favor of the approach described in this paper is also the fact that light does not penetrate very deep in tissue due to the small delivery-collection separation distance and thus does not reach tissue layers such as muscle and fat.

Finally, the method is not limited to steady-state diffuse reflectance and may be also useful for time domain or frequency domain applications of diffuse reflectance as well. It may also be adapted to model the spatial distribution of diffuse reflectance on the tissue surface. Those capabilities make the method even more attractive and open the way for an extended range of applications.

KEYWORDS

- **optical spectroscopy of skin**
- **diffuse reflectance**
- **Monte Carlo modeling**
- **optical ankle brachial index (O_{ABI})**

REFERENCES

1. Kaiser, H. Quantitation in Elemental Analysis. *Anal. Chem.* **1970**, *42* (2), 26A–59A.
2. Winefordner, J. D, Chapter 11, Trace Analysis: Spectroscopic Methods for Elements. In *Chemical Analysis*; First Edition; Wiley, 1976.

3. Mourant, J. R.; Freyer, J. P; Hielscher, A. H.; Eick, A. A.; Shen, D.; Johnson, T. M. Mechanisms of Light Scattering from Biological Cells Relevant to Noninvasive Optical-Tissue Diagnostics. *Appl. Opt.* **1998,** *37* (16), 3586–3593.
4. Cheng, H. H.; Ferradal, S. L.; Vyas, R.; Wigmore, D.; McDavitt, E.; Soul, J. S.; Franceschini, M. A.; Newburger, J. W.; Grant, P. E. Abnormalities in Cerebral Hemodynamics and Changes with Surgical Intervention in Neonates with Congenital Heart Disease. *J. Thorac. Cardiovasc. Surg.* **2020,** *159* (5), 2012–2021.
5. Kolpakov, A. V.; Moshkova, A. A.; Melikhova, E. V.; Sokolova, D. Y.; Muravskaya, N. P.; Samorodov, A. V.; Kopaneva, N. O.; Lukina, G. I.; Abramova, M. Y.; Mamatsashvili, V. G.; et al. Diffuse Reflectance Spectroscopy of the Oral Mucosa: In Vivo Experimental Validation of the Precancerous Lesions Early Detection Possibility. *Diagnostics* **2023,** *13* (9), 1633.
6. Roy, H. K.; Backman, V. Spectroscopic Applications in Gastrointestinal Endoscopy. *Clin. Gastroenterol. Hepatol.* **2012,** *10* (12), 1335–1341.
7. Martins, I. S.; Silva, H. F.; Lazareva, E. N.; Chernomyrdin, N. V.; Zaytsev, K. I.; Oliveira, L. M.; Tuchin, V. V.; Measurement of Tissue Optical Properties in a Wide Spectral Range: A Review [Invited]. *Biomed. Opt. Express* **2023,** *14*, 249–298.
8. Tuchin, V. V. In *Handbook of Optical Biomedical Diagnostics*, Second Edition; Tuchin, V. V., Ed.; SPIE Press: Bellingham, Washington, USA, 2016.
9. Jacques, S. L. Tutorial on Monte Carlo Simulation of Photon Transport in Biological Tissues. *Biomed. Opt. Express* **2023,** *14* (2), 559–576.
10. Zonios, G.; Dimou, A. Modeling Diffuse Reflectance from Semi-Infinite Turbid Media: Application to the Study of Skin Optical Properties. *Opt. Express* **2006,** *14*, 8661–8674.
11. Yudovsky, D.; Pilon, L. Retrieving Skin Properties from *in vivo* Spectral Reflectance Measurements. *J. Biophoton.* **2011**, *4*, 305–314.
12. Kim, O.; McMurdy, J.; Lines, C.; Duffy, S.; Crawford G.; Alber, M. Reflectance Spectrometry of Normal and Bruised Human Skins: Experiments and Modeling. *Physiol. Meas.* **2012,** *33* (2), 159–175.
13. Salomatina, E.; Jiang, B.; Novak, J.; Yaroslavsky, A. N. Optical Properties of Normal and Cancerous Human Skin in the Visible and Near-Infrared Spectral Range, *J. Biomed. Opt.* **2006,** *11* (6), 064026.
14. Kollias, N.; Seo, I.; Bargo, P. R. Interpreting Diffuse Reflectance for *in vivo* Skin Reactions in Terms of Chromophores. *J. Biophoton.* **2010,** *3*, 15–24.
15. Zonios, G.; Perelman, L. T.; Backman, V.; Manoharan, R.; Fitzmaurice, M.; Van Dam, J.; Feld, M. S. Diffuse Reflectance Spectroscopy of Human Adenomatous Colon Polyps in vivo. *Appl. Opt.* **1999,** *38*, 6628–6637.
16. Lisenko, S. A.; Kugeiko, M. M. Contactless Diagnostics of Biophysical Parameters of Skin and Blood on the Basis of Approximating Functions for Radiation Fluxes Scattered by Skin. *Quantum Electron.* **2014,** *44* (3), 252–258.
17. Feynman, R. P.; Hibbs, A. R. In *Quantum Mechanics and Path Integrals*, 1st Ed.; McGraw-Hill College, 1965.
18. Perelman, L. T.; Wu, J.; Itzkan, I.; Feld, M. S. Photon Migration in Turbid Media Using Path Integrals. *Phys. Rev. Lett.* **1994,** *72* (9), 1341–1344.
19. Perelman, L. T.; Wu, J.; Wang, Y.; Itzkan, I.; Dasari, R. R.; Feld, M. S. Time-Dependent Photon Migration using Path Integrals. *Phys. Rev. E* **1995,** *51* (6), 6134–6141.

20. Niemz, M. H. In *Laser-Tissue Interactions. Fundamentals and Applications*, 4th Edition; Springer Nature: Switzerland AG, 2019.
21. Allen, J. Photoplethysmography and its Application in Clinical Physiological Measurement. *Physiol. Meas.* **2007,** *28*, R1–R39.
22. Scott, P. S. J. Optical Absorption and Emission Data [Online]. http://omlc.ogi.edu
23. Roggan, A.; Friebel, M.; Doerschel, K.; Hahn, A.; Mueller G. J. Optical Properties of Circulating Human Blood in the Wavelength Range 400–2500 nm. *J. Biomed. Opt.* **1999,** *4* (1), 36–46.
24. Lisenko, S. A.; Kugeiko, M. M.; Lisenkova, A. M. Noninvasive Determination of Spectral Depth of Light Penetration into Skin. *Opt. Spectrosc.* **2013,** *115*, 779–785.
25. Patterson, M. S.; Chance, B.; Wilson, B. C. Time Resolved Reflectance and Transmittance for the Non-invasive Measurement of Tissue Optical Properties. *Appl. Opt.* **1989,** *28* (12), 2331–2336.
26. Feng, S.; Zeng, F. A.; Chance, B. Photon Migration in the Presence of a Single Defect: A Perturbation Analysis, *Appl. Opt.* **1995,** *34* (19), 3826–3837.
27. Cui, W.; Kumar, C.; Chance B. Experimental Study of Migration Depth for the Photons Measured at Sample Surface, in Time-Resolved Spectroscopy and Imaging of Tissues. *Proc. SPIE* **1991,** *1431*, 180–191.
28. Marti, D.; Aasbjerg, R. N. N.; Andersen, P. E. E.; Hansen A. K. K. MCmatlab: An Open-Source, User-Friendly, MATLAB-Integrated Three-Dimensional Monte Carlo Light Transport Solver with Heat Diffusion and Tissue Damage, *J. Biomed. Opt.* **2018,** *23* (12), 121622.
29. Lisenko, S. A.; Kugeiko, M. M. A Method of Online Quantitative Interpretation of Diffuse Reflection Profiles of Biological Tissues. *Opt. Spectrosc.* **2013,** *114*, 251–259.
30. Jacques, S. L. Skin Optics, Oregon Medical Laser Center News, 1998. http://omlc.org/news/jan98/skinoptics.html
31. Maeda, T.; Arakawa, N.; Takahashi M.; Aizu, Y. Monte Carlo Simulation of Spectral Reflectance using a Multilayered Skin Tissue Model. *Opt. Rev.* **2010,** *17* (3), 223–229.
32. Jaques, S. L. Optical Properties of Biological Tissues: A Review. *Phys. Med. Biol.* **2013,** *58* (11), R37–R61.
33. Lister, T.; Wright, P. A.; Chappell, P. H. Optical Properties of Human Skin. *J. Biomed. Opt.* **2012,** *17* (9), 90901-1.
34. Verkruysse, W.; Lucassen, G. W.; de Boer, J. F.; Smithies, D. J.; Nelson, J. S.; van Gemert, M. J. Modelling Light Distributions of Homogeneous Versus Discrete Absorbers in Light Irradiated Turbid Media. *Phys. Med. Biol.* **1997,** *42* (1), 51–65.
35. Prahl, S. A.; van Gemert, M. J. C.; Welch, A. J. Determining the Optical Properties of Turbid Media by using the Adding–Doubling Method. *Appl. Opt.* **1993,** *32*, 559–568.
36. Groenhuis, R. A. J.; Ferwerda, H. A.; Ten Bosch, J. J. Scattering and Absorption of Turbid Materials Determined from Reflection Measurements. 1: Theory. *Appl. Opt.* **1989,** *22*, 2456–2462.
37. Farrell, T. J.; Patterson, M. S.; Wilson, B. A. Diffusion Theory Model of Spatially Resolved, Steady State Diffuse Reflectance for the Noninvasive Determination of Tissue Optical Properties in vivo. *Med. Phys.* **1992,** *19* (4), 879–888.
38. Lim, L.; Nichols, B.; Rajaram, N.; Tunnell, J. W. Probe Pressure Effects on Human Skin Diffuse Reflectance and Fluorescence Spectroscopy Measurements. *J. Biomed. Opt.* **2011,** *16* (1), 011012.

39. Grabovskis, A.; Marcinkevics, Z.; Rubins, U.; Kviesis-Kipge, E. Effect of Probe Contact Pressure on the Photoplethysmographic Assessment of Conduit Artery Stiffness. *J. Biomed. Opt.* **2013,** *18* (2), 027004.
40. Shumkova, D. B.; Levchenko, A. E. In *Special Fiber Light Guides*, Perm Publishing House: National Research Polytechnic University, Perm, Russian, 2011; p 178. ISBN 978-5-398-00743-5
41. Ankle Brachial Index: Quick Reference Guide for Clinicians. *J. Wound Ostomy Continence Nurs.* **2012,** *39* (2S), S21–S29.

CHAPTER 18

External-Stimuli-Activated Single and Multi-Drug Delivery Systems Based on Liquid Crystal Micro-Containers

KETEVAN CHUBINIDZE[1], GIA PETRIASHVILI[1], TAMARA TATRISHVILI[2], ELENE KALANDIA[1], IRINE KHUBUA[3], TAMAR MAKHARADZE[1], MARIAM CHUBINIDZE[4], and ANA PETRIASHVILI[1]

[1]*Institute of Cybernetics of Georgian Technical University, Tbilisi, Georgia*

[2]*Department of Macromolecular Chemistry, Tbilisi State University, Tbilisi, Georgia*

[3]*Caucasus Medical Centre, Tbilisi, Georgia*

[4]*Jo Ann University Hospital, Tbilisi, Georgia*

ABSTRACT

We demonstrate novel, light, pH, and temperature-controlled drug delivery systems based on photochromic, pH-sensitive dye-doped liquid crystal (LC) microspheres and the temperature-dependent phase transition in LCs. Experimental results showed that when exposed to UV light, pH, and temperature, the photo- and thermochromic and pH-responsive molecules located inside the LC microspheres undergo interconversion of hydrophobic, oil-soluble, nonpolar state to the hydrophilic, water-soluble, highly polar condition. Isomerization caused by light, pH, and temperature destabilizes the LC-water interface, stimulates their translocation across the LC-water barrier, and leads to uniform distribution over an aqueous environment. The proposed modality may be considered as an advanced direction for the

Advances in Engineering Materials: Preparation for Sustainable Process Development. Tamara Tatrishvili and A. K. Haghi (Eds.)

photo-stimulated, pH, and temperature-sensitive drug delivery systems that offer controlled delivery and release capabilities for a wide range of drugs into the body at the suitable time and the desired "sites" to combat various diseases, including cancer.

18.1 INTRODUCTION

Despite clinical breakthroughs in cancer therapy, drug resistance inevitably emerges in cancer treatment. Multidrug therapy by mixing individual drugs based on traditional drug delivery has proven to be a simple and efficient approach to providing systematic therapy to treat complex health conditions.[1,2] Using the traditional methods of drug administration, about 99% of chemotherapy drugs do not reach the cancerous cells by employing conventional methods in cancer treatment. Colloidal drug carrier systems, such as micellar solutions, vesicles, and liquid crystal dispersions, show great promise. Controlled drug delivery systems (CDDS) are becoming increasingly important compared to traditional forms of drugs. Delivering drugs at a controlled rate and in targeted placement accumulates therapeutic compounds in the target site and, consequently, undesirable side effects can be minimized. CDDS have shown a steep rise in tumor therapy application with the constantly evolving nanotechnology. Nanocarriers were widely used in oncotherapy as the characteristics of tumor targets, drug-controlled release, or blood-long circulation.[3] Two approaches can be adopted in designing stimuli-responsive CDDS. Two methods can be taken when developing a stimulus-responsive CDDS. In one approach, endogenous stimuli (intrinsic stimuli/biological stimuli), which are unique to the affected tissue, can increase the specificity of drug action. These include pH and varying levels of glutathione concentration due to oxidative stress and enzyme overexpression.[4,5] Regarding the second approach, the physical stimuli are applied externally to the target tissue after delivery of specific drug-loaded nanocarriers. These external triggers integrate light, magnetic and electrical fields, temperature, and ultrasound.[6–9]

Liquid crystals are a unique state of matter that exhibits properties of both liquids and crystalline solids. They have characteristics of liquids because their molecules can flow and move like a fluid, but they also have properties of crystals because their molecules can align in ordered patterns. The molecules in a liquid crystal are typically elongated and have a rigid core with flexible ends. These molecules have a mesogenic structure, which means they can arrange themselves in an ordered manner. The specific arrangement depends on the liquid crystal type and its temperature, as shown in Figure 18.1.

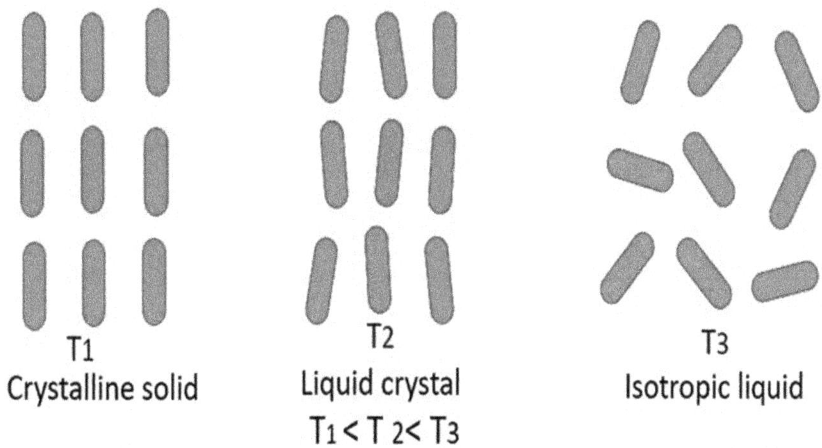

FIGURE 18.1 Temperature-dependent orientation of liquid crystal molecules.

Liquid crystalline materials show variable and uncial optical properties controlled by external forces, such as light, temperature, electric, magnetic, acoustic fields, and pressure. Due to their self-assembly ability, high sensitivity, chemical diversity, and tunability, stimulus-responsive liquid crystal biosensors are reported to have bright prospects and advantages in biological applications.[10] Among the liquid crystal phases, the cholesteric liquid crystals (CLCs) have a particular symmetry, which can be regarded as self-assembled one-dimensional photonic crystals. The molecules in CLCs are self-assembled in a supramolecular helical structure, which results in periodic changes in the refractive index within the material. Due to the inversion symmetry of molecules, the periodicity of the structure is equal to half the pitch of the helix, p. When light propagates along the helical axes, it experiences Bragg reflection at $\lambda_0 = np$, where λ_0 is the wavelength of maximum reflection and n is the average value of the refractive indices, defined as $n = (n_e + n_o)/2$, where n_o and n_e are the ordinary and extraordinary refractive index, respectively. As a result, the entire range of wavelengths does not propagate within the material, which is defined as a photonic band gap (PBG). The full PBG width at half maximum is $\Delta\lambda = p\Delta n$, where $\Delta n = n_e - n_o$ is the birefringence of the liquid crystal. The spectral position of the PBG is sensitive to internal and external factors such as electric and electromagnetic fields, temperature, and variations in local order.[11] Here we note, that liquid crystals consist of organic molecules, and the majority of viruses have liquid crystalline symmetry, as shown in Figures 18.2 and 18.3.

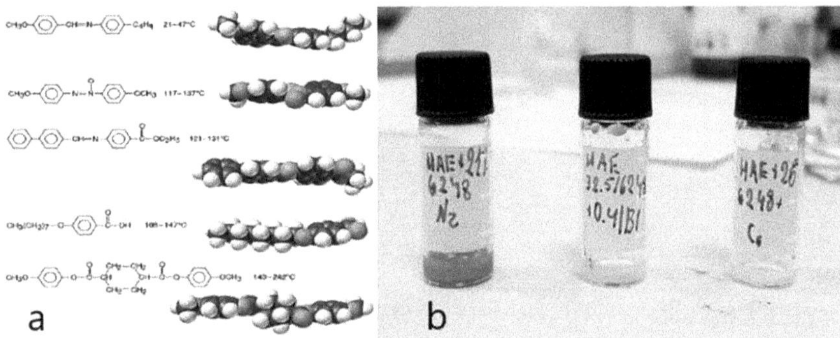

FIGURE 18.2 (a) Structural and 3D presentation of liquid and (b) vials with liquid crystals crystal molecules.

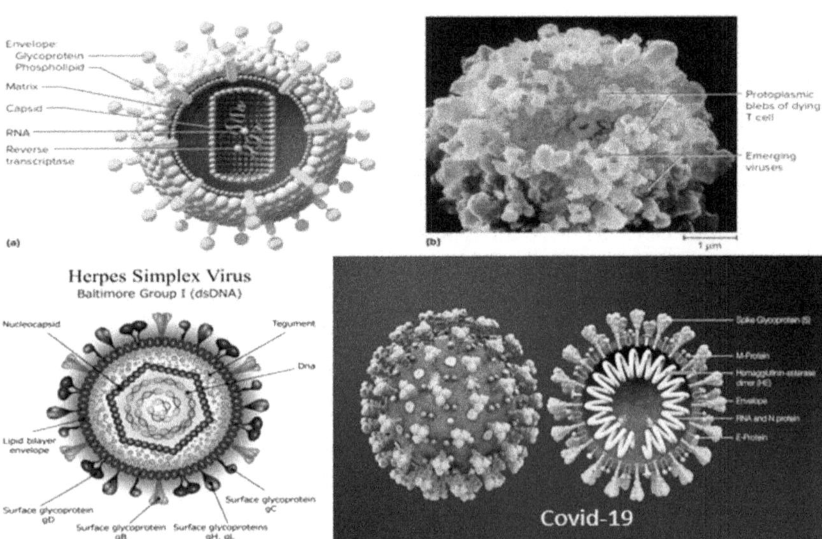

FIGURE 18.3 Many viruses have liquid crystalline symmetry, especially those belonging to human immunodeficiency viruses.

18.2 EXPERIMENTAL SECTION

18.2.1 PREPARATION OF THE CHOLESTERIC LIQUID CRYSTALS

The CLC mixtures were prepared by mixing the initial nematic matrices and optically active dopants, which were received from Merck. The prepared mixtures were stirred in the isotropic phases at 80–100°C for 15–20 min to make the constituents uniformly mixed. Table 18.1 shows the CLC mixtures' sensitivity upon exposure to pH, photo, and temperature, respectively.

TABLE 18.1 List of CLC Mixtures Sensitive to pH, Light, and Temperature Variations.

Liquid crystal mixture	pH-responsive	Photo-responsive	Temperature-responsive
76% BL006 + 24% MLC-6248	+		+
73% BL006 + 27% MLC-C6247	+		
76% BL090 + 24% MLC-C6248		+	+
75% BL090 + 25% ZLI-811		+	
72% BL090 + 28% ZLI-811	+		+
68% BL090 + 32% ZLI-811		+	
68.5% BL090 + 31.5% MLC-6248	+	+	
70% BL090 + 15% ZLI-811+ 15% MLC-6247	+	+	
68% BL090 + 17% ZLI-811+ 15% MLC-6247		+	+
70% BL090 + 20% ZLI-811+ 10% MLC-6247		+	+
67.5% BL006 + 22.5% ZLI-811+ 10% MLC-6247	+	+	+
95% [72.5% BL090 + 27.5% MLC-6248]	+		+
95% [76% BL090 + 24% MLC-6248]	+	+	
95% [73% BL090 + 27% MLC-6248]	+		+
95% [69% BL090 + 15% MLC-6247+17% ZLI-811]		+	+
95.5% [69% BL006 + 10% MLC-6247+22.5% ZLI-811]		+	
96% [66% BL006 + 10% MLC-6247+24% ZLI-811]	+	+	
96% [66.5% BL006 + 14% MLC-6247 + 20.5% ZLI-811]	+		+

TABLE 18.1 *(Continued)*

Liquid crystal mixture	pH-responsive	Photo-responsive	Temperature-responsive
92% [66.5% BL006 + 14% MLC-6247 + 20.5% ZLI-811]		+	+
75.4% MLC-6815 + 24.6% MLC-6247	+	+	+
77.4% MLC-6815 + 22.6% MLC-6247	+		+
68% MLC-6815 + 32% MLC-6247	+	+	
66% MLC-6815 + 34% MLC-6247		+	+
64% MLC-6815 + 36% MLC-6247	+	+	+
63% MLC-6815 + 37% MLC-6248			
74% MLC-6816 + 26% MLC-6248		+	+
73% MLC-6816 + 27% MLC-6247	+		
69.7% MLC-6816 + 30.3% MLC-6247	+	+	+
68.9% MLC-6816 + 31.1% MLC-6247	+		+
92% (73% MLC-6816 + 27% MLC-6247)		+	+
92% (70% MLC-6816 + 30% MLC-6247)		+	+
96% (68% MLC-6816 + 32% MLC-6247)		+	
97% (73.5% MLC-6816 + 26.5% MLC-6247)	+	+	
96% (68.9% MLC-6816 + 31.1% MLC-6247	+	+	+
96% (68.9% MLC-6816 + 31.1% MLC-6247)	+		+
95% (67.5% MLC-6816 + 23.5% MLC-6248)	+	+	

An optical microscope coupled with a fiber-optic spectrometer (Avaspec-2048, Avantes) with 1 nm accuracy and a high-resolution charge-coupled device (CCD) camera served as the imaging-based technique for the visual and graphical presentation of the light transmitted through the CLC and reflected from it. To form CLC–water emulsions, the mixtures of CLC and water were prepared with the following percentage in weight, shown in Table 18.1. To obtain a homogeneous distribution of CLC microspheres in the water-glycerol matrix and to stabilize the emulsion, glycerol has been added to the aqueous solution. The prepared mixtures were stirred at 600 rpm for 10 min at room temperature. Finally, a homogeneous distribution of CLC microspheres suspended in an aqueous matrix was obtained. The diameters of CLC microspheres were ranging from 10 to 15 μm, as shown in Figure 18.4.

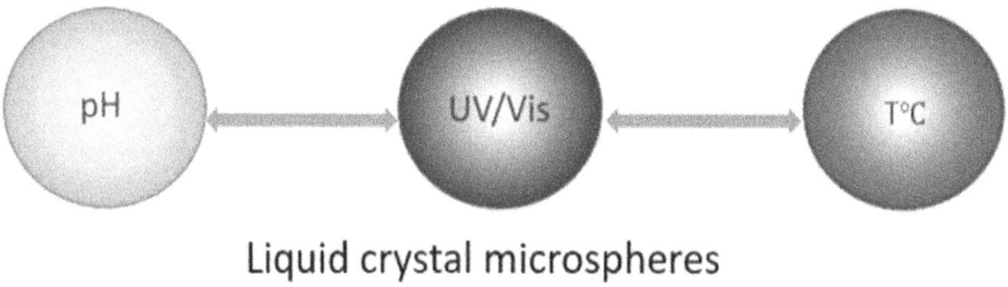

FIGURE 18.4 Schematic representation of CLC microspheres activated upon exposure to pH, UV/Vis, and temperature stimuli.

18.2.2 LIGHT-CONTROLLED DRUG TRANSPORTING MICROSPHERES BASED ON PHOTOSWITCHABLE MOLECULES DOPED LIQUID CRYSTAL MICROSPHERES

We have elaborated novel, light-controlled drug-transporting microcontainers based on the spiropyran (SP) emulsion formed by the immiscibility between the liquid crystal microspheres and a water environment, as shown in Figure 18.5.

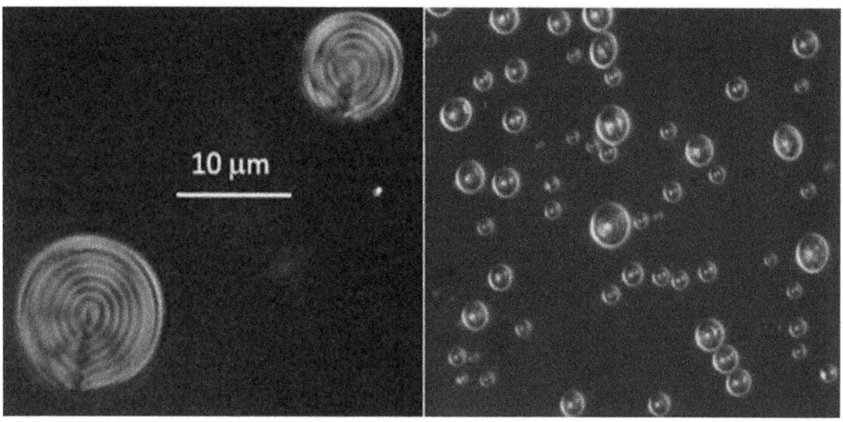

FIGURE 18.5 LC microspheres emulsified in the water environment.

The stable state SP is an uncolored closed molecular form that can transform to the merocyanine (MC) state upon UV/violet irradiation, which in turn destabilizes the lipid membrane at the nanostructured lipid carriers (NLC)-water interface so that the MC molecules can translocate into the aqueous environment, as shown in Figure 18.6.

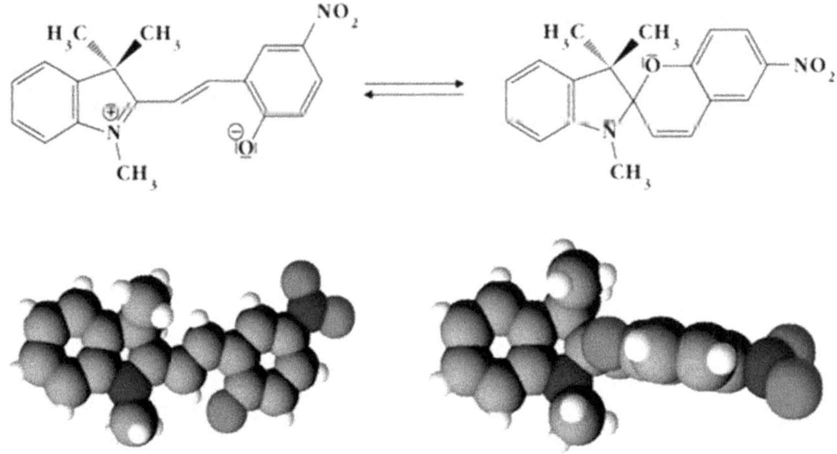

FIGURE 18.6 The light-stimulated transformation from SP to MC form.

As to our concept, SP molecules initially bind to specific therapeutic drug molecules, and this ensemble is trapped inside the NLC microdroplets. When exposed to UV radiation, which results in photoisomerization of SP–MC, the MC–drug molecule combination moves across the LC–water barrier and distributes uniformly throughout the extracellular environment.

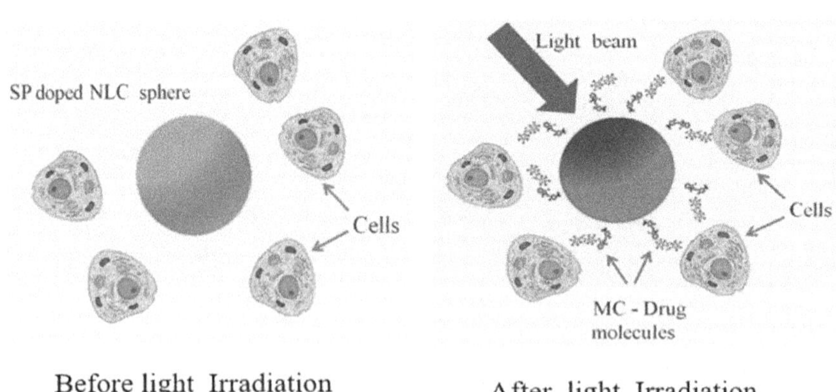

FIGURE 18.7 Schematic of light-activated drug release. SP-NLC microsphere before (left picture) and after (right picture) UV exposures.

18.2.3 TEMPERATURE-CONTROLLED DRUG DELIVERY

The strategy of temperature-controlled drug delivery is promising since the drug is self-released in those areas of the body where the temperature has elevated or there is a targeted increase in temperature in the preferred parts of the body at desired times. For this purpose, we have prepared and examined the temperature-responsive LC microspheres doped with therapeutic drugs, as shown in Figure 18.8.

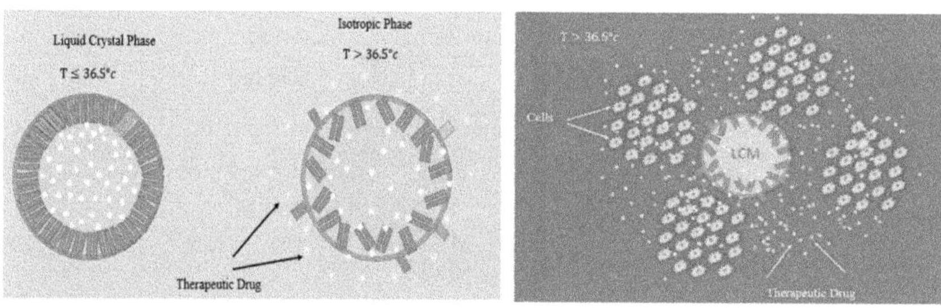

FIGURE 18.8 Schematic of temperature-controlled drug release. At low temperatures, the therapeutic drugs are located inside the CLC microspheres, whereas, at high temperatures, the drug particles are disseminated throughout the biological cells.

18.2.4 pH CONTROLLED DRUG DELIVERY MICRO CONTAINERS

pH-Controlled drug delivery systems are gaining importance as these systems because different organs, such as tissues and cellular parts, have distinct pH values, making pH a suitable stimulus for controlled drug release. For example, pH decrease is observed in most solid tumors, and drug-delivery systems responsive to the slightly acidic extracellular. Inflammation, tumors, and circulatory disorders locally disorder the acid–base balance in the body. These changes in pH value can be used, for example, to test the effectiveness of cancer treatments. However, until now, there has not been an imaging technique to make such changes visible in patients. We have elaborated a pH sensor that visualizes the pH values in a noninvasive and radiation-free way. In this experiment, we prepared an LC/water microemulsion, but in this case, the fluorescent dye Nile Red was doped inside the LC microspheres as the pH-sensitive one, as shown in Figure 18.9. This molecule smartly responds to changes in environmental pH, resulting in prompt movement through the LC-water interface and uniform distribution throughout the aquatic environment. In the proposed case, the infiltration of acid molecules inside microdroplets produces two effects: a change in the shape of the photonic bandgap and a shift in the dye absorption and fluorescence band spectral position, as shown in Figures 18.10 and 18.11.

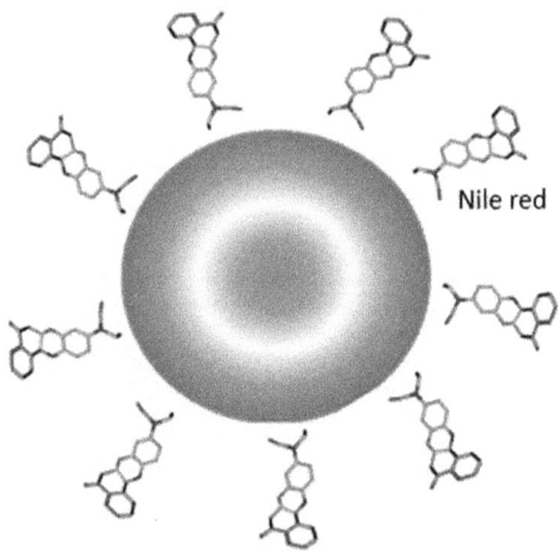

FIGURE 18.9 CLC microsphere surrounded with fluorescent dye Nile red.

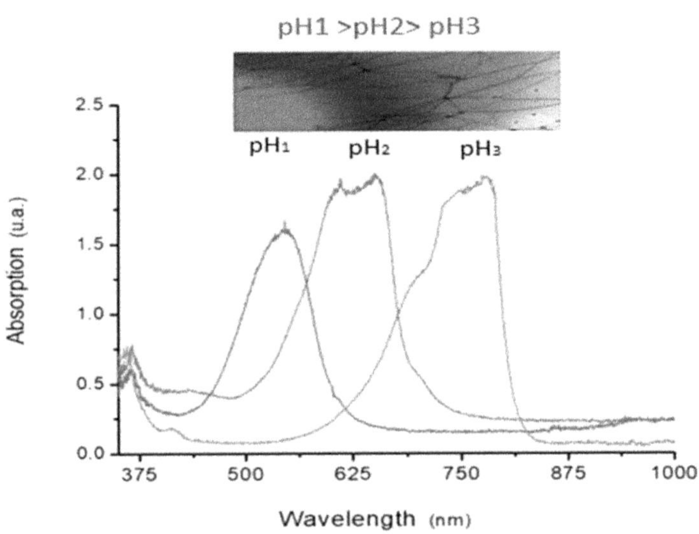

FIGURE 18.10 pH-dependent absorption shifting of FD dissolved in LC microspheres.

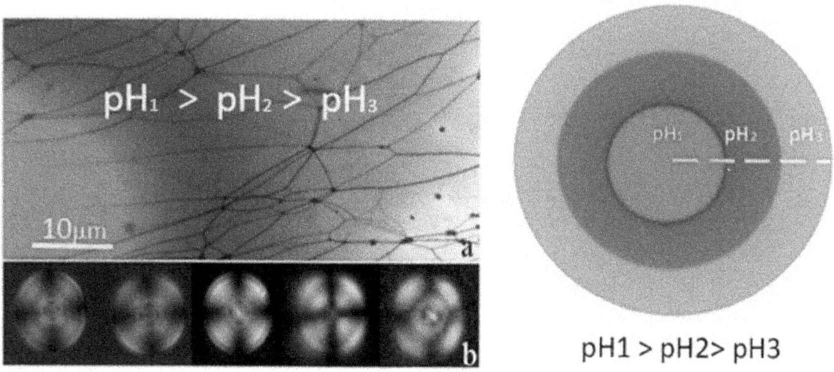

FIGURE 18.11 Distribution and gradient of lactic acid inside the LC micro droplet.

18.2.5 CONTROLLED MULTIDRUG DELIVERY PLATFORM

Despite the clinical breakthroughs emerging in cancer therapy, drug resistance inevitably appears in cancer treatments. The concept of multidrug therapy, by mixing individual drugs based on conventional drug delivery administration, has proven to be a simple and efficient approach to providing systematic therapy to treat complex health conditions. In this study, we propose a new concept of a multidrug delivery system with sequential release based on nanoparticle (NP)-doped liquid crystal microspheres as drug carriers. NPs have been incorporated into various materials, such as semiconductors, core-membrane composites, metals, and organic polymers. This research on micro/nano colloids loaded in LCs has inherent interest due to their potential application in modern scientific and technological directions.[12,13] In these experiments, we prepared three CLC mixtures and loaded them with Gold (a), Silver (b), and copper (c) nanoparticles, as shown in Figure 18.12.

To fabricate the multidrug delivery containers loaded with NPs, first, three CLC Droplets were mixed in a water–gelatin solution at 800 rpm for 20 min. Since CLC and gelatin-water solutions have no miscibility, this procedure allows the formation of CLC microspheres, the size of which depends on the stirring speed and duration of mixing, as shown in Figure 18.13.

Gelatin films consisting of three different NP-loaded liquid crystal microspheres were prepared using a method for microencapsulation. Controlling all the parameters of microencapsulation, such as stirring speed, temperature, and concentration ratio of the constituents, the microcapsules with desired sizes, packing, and spatial distributions have been obtained, as shown in Figure 18.14.

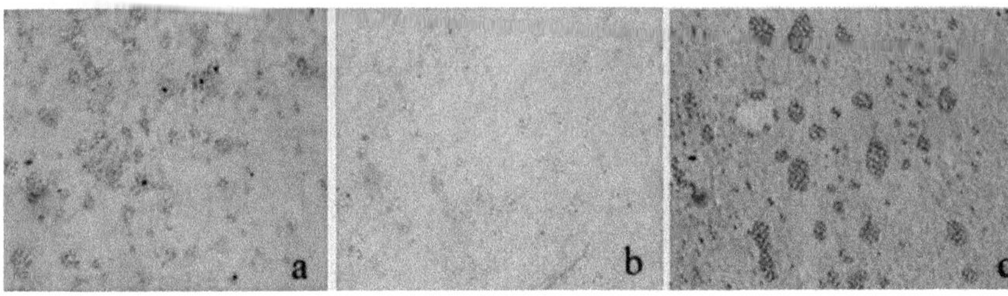

FIGURE 18.12 Prepared three cholesteric liquid crystal mixtures and loaded them with gold (a), silver (b), and copper (c) nanoparticles.

FIGURE 18.13 Preparation of cholesteric liquid crystal microspheres.

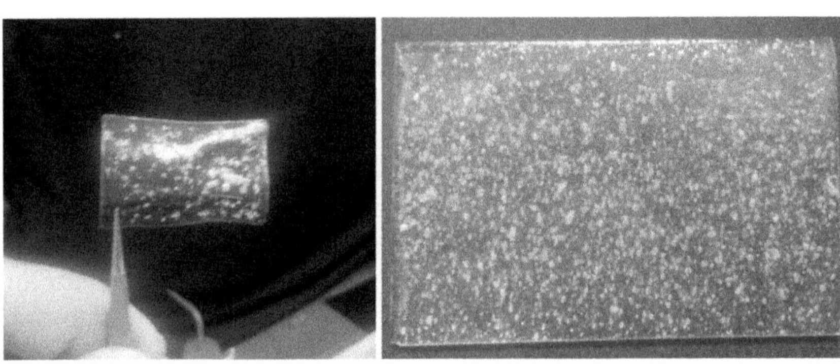

FIGURE 18.14 Liquid crystal microspheres encapsulated in Gelatin matrix.

The microscopic images of Gelatin film consist of three different kinds of liquid crystal microspheres as drug carriers that provide two functions:

multidrug delivery and sequential drug release. The average size of microspheres is 20 μm. Figures 18.15 and 18.16 show the NP-doped LCs microspheres using optical-polarizing and fluorescent microscopes, respectively.

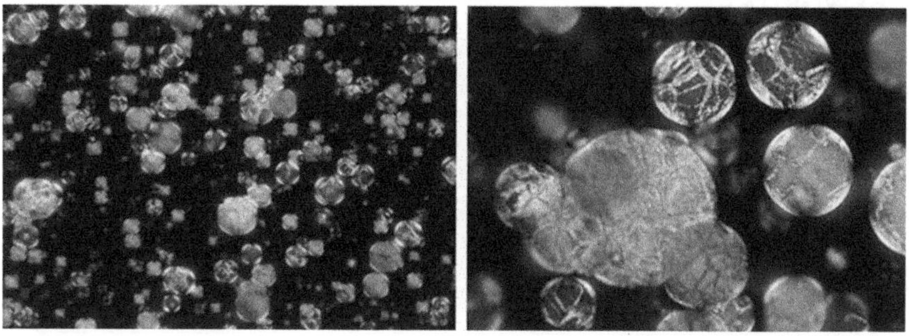

FIGURE 18.15 Nanoparticle-doped LCs microspheres as they look under an optical-polarizing microscope.

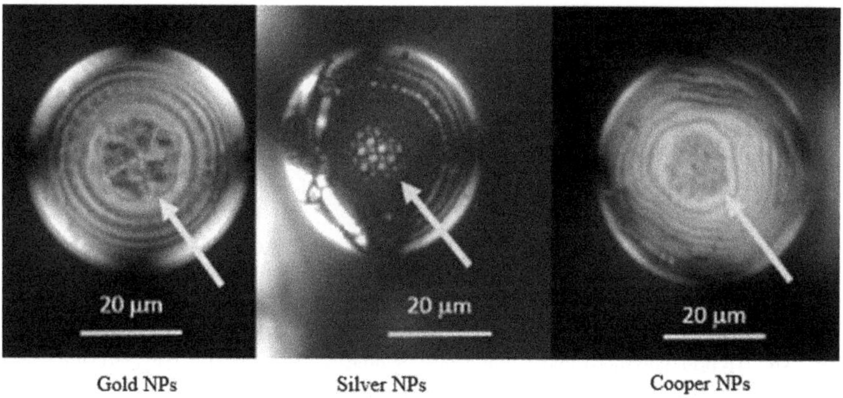

FIGURE 18.16 Nanoparticle-doped LC microspheres as they look under a fluorescent microscope with high magnification.

18.3 CONCLUSION

In this study, we propose a new concept of a multidrug delivery system with sequential release based on NP-doped liquid crystal microspheres as drug carriers. The proposed strategy can be considered as a new platform for the photo, pH, and temperature-stimulated drug delivery systems that offer the possibilities of the controlled delivery and release of a wide variety of drugs into the body at suitable times and desired sites, fighting different kinds of diseases including cancer diseases.

KEYWORDS

- **drug delivery systems**
- **liquid crystals**
- **pH-sensitive**
- **photo-sensitive**
- **temperature-sensitive**

REFERENCES

1. Gao, J.; Karp, J. M.; Langer, R.; Joshi, N. The Future of Drug Delivery. *Chem. Mater.* **2023,** *35*, 359–363.
2. Chirakadze, A.; Mitagvaria, N.; Jishiashvili, D.; Petriashvili, G.; Dvali, N.; Shiolashvili, Z.; Chubinidze, K.; Makhatadze, N.; Jishiashvili, A.; Buachidze, Z.; Khomeriki, Ir. Microwave Synthesis, Characterization and Testing of Acute Toxicity of Boron Nitride Nanoparticles by Monitoring of Behavioral and Physiological Parameters. *Bull. Georg. Natl. Acad. Sci.* **2012,** *15*, 120–125.
3. Jia, R.; Teng, L.; Gao, L.; Su, T.; Fu, L.; Qiu, Zh.; Bi, Y. Advances in Multiple Stimuli-Responsive Drug-Delivery Systems for Cancer Therapy. *Int. J. Nanomed.* **2021,** *16*, 1525–1551.
4. Li, J.; Ma, Y. J.; Wang, Y.; Chen, B. Z.; Guo, X. D.; Zhang, C. Y. Dual Redox/pH-Responsive Hybrid Polymer-Lipid Composites: Synthesis, Preparation, Characterization and Application in Drug Delivery with Enhanced Therapeutic Efficacy. *Chem. Eng. J.* **2018,** *341*, 450–461.
5. Raza, A.; Rasheed, T.; Nabeel, F.; Hayat, U.; Bilal, M.; Iqbal Hafiz, M. N. Endogenous and Exogenous Stimuli-Responsive Drug Delivery Systems for Programmed Site-Specific Release. *Molecules* **2019,** *24*, 1117.
6. Rasheed, T.; Bilal, M.; Abu-Thabit, N. Y.; Iqbal, H. M. The Smart Chemistry of Stimuli-Responsive Polymeric Carriers for Target Drug Delivery Applications. In *Stimuli Responsive Polymeric Nanocarriers for Drug Delivery Applications*; Elsevier: Amsterdam, The Netherlands; Woodhead Publishing: Cambridge, UK, 2018; vol *1*, pp 61–99.
7. Chubinidze, K.; Partsvania, B.; Devadze, L.; Zurabishvili, T.; Sepashvili, N.; Petriashvili, G.; Chubinidze, M. Gold Nanoparticle Conjugated Organic Dye Nanocomposite Based

Photostimulated Luminescent Enhancement and Its Application in Nanomedicine. *Am. J. Nano Res. Appl.* **2017,** *5*, 42–47.
8. Chubinidze, K.; Partsvania, B.; Khuskivadze, A.; Burnadze, P.; Petriashvili, G.; Dzidziguri, D.; Mukbaniani, O. Modeling of Calmodulin-Mediated Processes in Tissues Using Calmodulin -Functionalized Gold Nanoparticles and Fluorescent Dyes. *Mater. Technol.* **2020,** *54*, 211–214.
9. Petriashvili, G.; Devadze, L.; Zurabishvili, T.; Sepashvili, N.; Chubinidze, K. Light Controlled Drug Delivery Containers Based on Spiropyran Doped Liquid Crystal Micro Spheres. *Biomed. Opt. Express* **2016**, *7*, 442–447.
10. Rastogi, A.; Singh, A.; Naik, K.; Mishra, A.; Chaudhary, S.; Manohar, R.; Parmar, A. S. A Systemic Review on Liquid Crystals, Nanoformulations and its Application for Detection and Treatment of SARS – CoV- 2 (COVID – 19). *J. Mol. Liq.* **2022,** *362*, 119795.
11. Petriashvili, G.; Bruno, M. D. L.; De Santo, M. P.; Barberi, R. Temperature-Tunable Lasing from Dye-Doped Chiral Microdroplets Encapsulated in a Thin Polymeric Film. *Beilstein J. Nanotechnol.* **2018,** *9*, 79–383.
12. Chenthamara, D.; Subramaniam, S.; Ramakrishnan, S. J.; Krishnaswamy, S.; Essa, M. M.; Lin, F. H.; Qoronfleh, M. W. Therapeutic Efficacy of Nanoparticles and Routes of Administration. *Biomater. Res.* **2019.** https://doi.org/10.1186/s40824-019-0166-x
13. Seaton, A.; Tran, L.; Aitken, R.; Donaldson, K. Nanoparticles, Human Health Hazard and Regulation. *J. R. Soc. Interface* **2010**, *7*, 119–S129.

CHAPTER 19

Chitosan/Sodium Alginate-Based Hydrogels: Preparation and Evaluation of the Influence on Biophysical Skin Parameters

JUSTYNA KOZLOWSKA, JULIA RYPINSKA, and KLAUDIA BRZEZINSKA

Department of Biomaterials and Cosmetic Chemistry, Faculty of Chemistry, Nicolaus Copernicus University in Torun, Poland

ABSTRACT

In this study, hydrogel based on the blend of chitosan and sodium alginate with the addition of *Centella asiatica*extract, royal jelly, panthenol, and glycerol has been examined for potential cosmetic or dermatological applications. Skin parameters measurements included trans-epidermal water loss (TEWL), erythema, stratum corneum hydration, skin oiliness, stiffness, and its elasticity using Courage–Khazaka multiprobe adapter (MPA) skin probes (Tewameter, colorimeter, Corneometer, Sebumeter, and Indendometer, respectively). To compare the effect of the obtained hydrogel on the skin, a hydrogel without adding active ingredients was prepared as a control sample. The skin condition was improved after applying the obtained hydrogels, especially in the case of hydrogel containing *C. asiatica* extract, panthenol, glycerol, and royal jelly. The designed hydrogels can be envisioned for creating new cosmetics for protective skin in cosmetic masks and potential hydrogel vehicles in dermatological formulations for pharmaceutical delivery.

Advances in Engineering Materials: Preparation for Sustainable Process Development. Tamara Tatrishvili and A. K. Haghi (Eds.)

19.1 INTRODUCTION

The skin is the human body's largest organ, consisting of three superimposed layers: the epidermis, dermis, and subcutaneous tissue. It contains hair follicles, sebaceous and sweat glands, blood vessels, and nerve endings.[1,2] The skin is a barrier for the human body that separates the internal and external environments, thus protecting against the effects of physical, chemical, and biological factors. The passive functions of the skin, resulting in its structure, include protection against cold, heat, radiation, mechanical injuries, and the penetration of pathogenic microorganisms. This organ also actively participates in thermoregulation, immune reactions, perception of stimuli from the external environment, metabolism of proteins, carbohydrates, and lipids, regulation of water and electrolyte balance, resorption of individual chemical compounds, and secretory activities.[1,3]

The stratum corneum is the outermost and thinnest layer of the skin. Because it borders the external environment, it mainly plays a protective role, and its average thickness is about 0.1 mm. This value may vary depending on the area of the body.[4] The skin is exposed to mechanical damage every day. Abrasion occurs when the epidermis layer of the skin is damaged. Damage skin, which can affect anyone, is caused by rubbing one skin area against another or clothing rubbing against it. This is a common problem that leads to pain and wounds, and ignoring even seemingly harmless skin redness may cause infection.[5] Skin abrasions occur more often when people are exposed to excessive sweating. Moreover, heat and humidity promote abrasions, which is why, especially in summer, people are prone to them. This problem also affects obese people when folds of skin rub against each other, and a moist environment is created between them. For this reason, proper skin care is very important to strengthen its condition, epidermal barrier, hydration, and mechanical properties. One of the solutions is hydrogel materials intended for skin applications, which, depending on the active substances or drugs contained, can prevent or alleviate many skin problems.[6]

The aim of our study was to develop and evaluate chitosan/sodium alginate-based hydrogel formulations enriched with active ingredients such as *Centella asiatica* extract, panthenol, glycerol, and royal jelly. Chitosan is the partially deacetylated form of chitin, where a fraction of the N-acetyl-D-glucosamine has been converted into D-glucosamine

(Figure 19.1a).[7] There are many sources of chitin in nature; this polymer can be found in the shells of crustaceans, such as shrimp, lobsters, crabs, insects (e.g., houseflies and silkworms), mollusks (e.g., octopuses and clams), and many others.[8] Due to biocompatibility, biodegradability, low immunogenicity, antimicrobial activity, non-toxicity, low antigenicity, low cost, and easy availability, chitosan has many industry applications, such as biomedical, cosmetic, and food.[9–11] Alginates are naturally occurring polysaccharides with excellent biocompatibility, gel-forming ability, non-toxicity, biodegradability, thickening properties, low production cost, and wide availability.[12–15] Sodium alginate is a linear copolymer of β-D-mannuronic acid (M) residues linked with α-L-guluronic acid (G) with 1,4-glycosidic bonds (Figure 19.1b). Sodium alginate, a hydrogel form of alginate, is used in various biomedical fields, such as drug delivery, wound healing, and tissue engineering, as well as in the food, pharmaceutics, and cosmetic industries.[16,17]

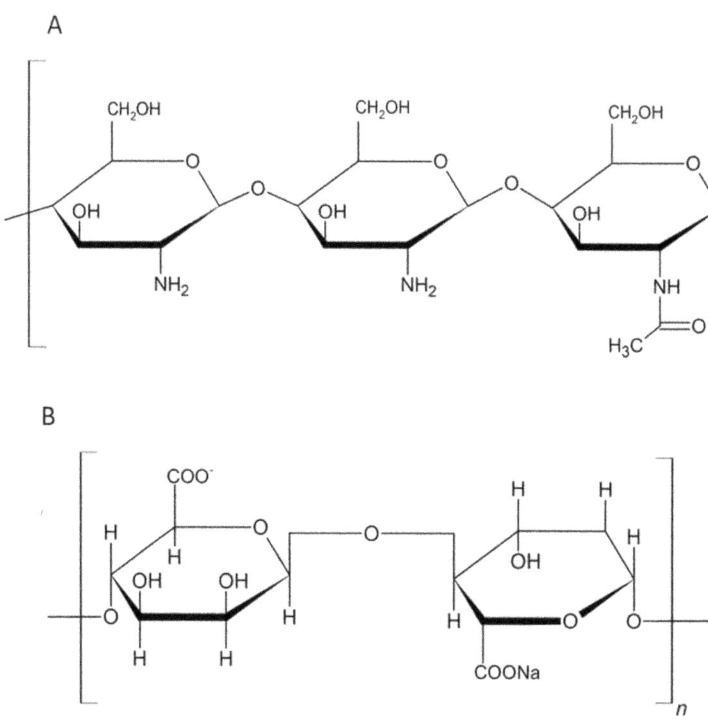

FIGURE 19.1 The structure of chitosan (a) and sodium alginate (b).

C. asiatica extract improves skin firmness, elasticity, and hydration. As a result, this ingredient has a regenerative effect, accelerates wound healing, and prevents scarring. It also has anti-inflammatory and soothing effects. The extract is often used in products to help treat atopic skin lesions and psoriasis.[18] Panthenol is a hydrophilic moisturizer (active ingredient). It has a moisturizing, anti-inflammatory, soothing effect and accelerates the epidermisFigures regeneration.[19] Glycerol is a hydrophilic moisturizer and penetration promoter.[20] Royal jelly may have an antibacterial effect on wounds and other inflammatory skin conditions like acne. It can promote firmer and smoother skin by boosting collagen production.[21] The active substances in our study were chosen to perfectly care for the skin, prevent damage to the epidermis, and soothe possible lesions.

The main objective of our study was to develop, characterize, and evaluate the effectiveness of a dermatological product based on chitosan, sodium alginate, *C. asiatica* extract, panthenol, glycerol, and royal jelly that should prevent skin abrasion and provides long-lasting hydrating and soothing effects.

19.2 EXPERIMENTAL METHODS AND MATERIALS

19.2.1 MATERIALS

Chitosan (low molecular weight) was purchased from Sigma–Aldrich (Poznan, Poland). Sodium alginate was acquired from BÜCHI Labortechnik AG (Flawil, Switzerland). *C. asiatica* extract, 75% panthenol hydrous solution, and vegetable glycerol were provided by ZrobSobieKrem (Prochowice, Poland). Royal jelly was purchased from Calaya (Zlotow, Poland).

19.2.2 PREPARATION OF POLYMERIC HYDROGELS

Hydrogel based on chitosan and sodium alginate was prepared by the dissolution-evaporation method. Firstly, 0.4% (m/v) sodium alginate water solution and a 3.6% (m/v) chitosan solution in 2% (v/v) acetic acid were prepared. Then, the alginate solution was added slowly to the chitosan solution while stirring with a magnetic stirrer at 45°C until a homogeneous mixture was obtained. The suitable volumes of solutions were mixed and the weight ratio of the sodium alginate/chitosan was 1:9. The second hydrogel contained the same amounts of the mentioned polymers, but 50 mL of a mixture of polymers was supplemented with 2 g of active ingredients (0.3 g of vegetable glycerol, 0.7 g of *C. asiatica* extract, 0.4 g of 75% panthenol hydrous solution, 0.6 g of royal jelly). The prepared mixtures were homogenized at 900 rpm for 6 min. Then, an ultrasonic bath was used to remove air bubbles from the mixtures until degassing. Thin films were obtained by casting a solution on glass plates (7 cm diameter), and the samples were dried in an oven at 60°C for 24 h.

19.2.3 SKIN EXAMINATION

The skin moisture, color, oiliness, elasticity, and TEWL index after application of the chitosan and sodium alginate-based hydrogels were examined using the following probes of the Courage–Khazaka MPA apparatus: Corneometer® CM 825, Colorimeter CL 400, Sebumeter® SM 815, Indentometer IDM 800 and Tewameter TM Hex, respectively. After informed oral consent, the measurements were managed on the forearm skin of three women (of age 22 years) with normal skin. First, skin parameters were measured before the hydrogel was applied to it (t_0). Then, the films were immersed in water for 1 min. Next, they were put to the skin and removed after 15 min. The influence of the obtained hydrogels on the biophysical skin parameters was measured immediately after removing the sample from the skin (t_1), 30 (t_2), 60 (t_3), and 120 min (t_4) after sample application. Moreover, the skin area where hydrogels were not applied was identified as the control area (CA). In contrast, the areas where hydrogels were applied were identified as the test area (TA). The measurement conditions were constant—temperature was in the 22–25°C range, and relative humidity was in the range 40–60%. Five measurements were performed with a given probe at each time point, and the results presented are the average value. The study was conducted by the Declaration of Helsinki, and the protocol was approved by the Ethics Committee of KB 67/2021.

19.3 RESULTS AND DISCUSSION

19.3.1 HYDROGELFIGURE S MORPHOLOGY

A photograph of the prepared polymer hydrogels is shown in Figure 19.2, while pictures of the polymer hydrogels applied to the skin of the forearm are presented in Figures 19.3 and 19.4.

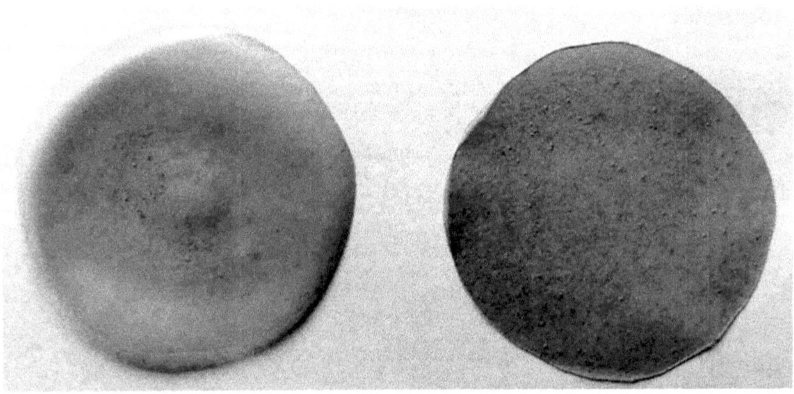

FIGURE 19.2 Basic hydrogel (on the left) and hydrogel with active ingredients—*Centella asiatica* extract, panthenol, royal jelly, and glycerol (on the right).

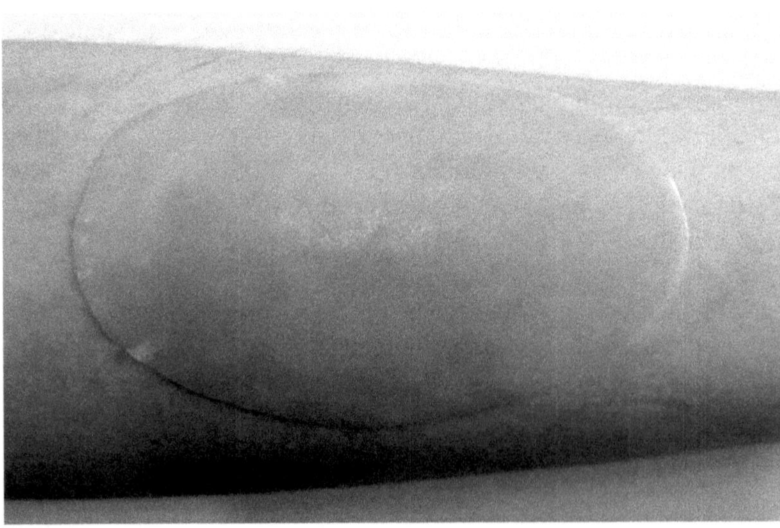

FIGURE 19.3 Basic hydrogel on the forearm.

The hydrogel containing the active ingredients was darker and more flexible than the basic hydrogel.

19.3.2 BIOPHYSICAL SKIN PARAMETERS ASSAYS

19.3.2.1 SKIN COLOR

The skin color before and after applying samples on the skin was determined using CIELAB color space. The CIELAB system represents the lightness of the color (L*), red/green intensity (a*), and yellow/blue intensity (b*). Therefore, the changes of a* values reveal skin redness and erythema. Moreover, the Individual Typology Angle (ITA°), defining an individualFigures skin color classification, is automatically calculated during measurement. As can be seen by analyzing the measurement data presented in Table 19.1, the parameters determining the brightness (L*), color (ITA°), and redness (a*) of the skin did not change significantly after both hydrogel applications. This means that our hydrogels do not irritate the skin or cause inflammatory reactions.

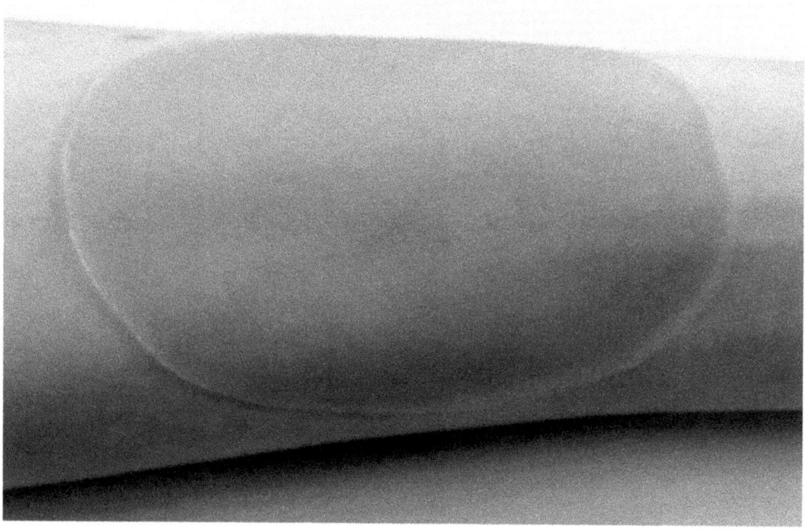

FIGURE 19.4 Hydrogel with active ingredients (*Centella asiatica* extract, panthenol, royal jelly, and glycerol) on the forearm.

TABLE 19.1 Skin Color after Application of Hydrogel Films.

Type of hydrogel	t_0	t_1	t_2	t_3	t_4
			L*		
Basic hydrogel	74.20 ± 0.56	73.13 ± 0.58	73.42 ± 0.34	71.89 ± 0.44	72.34 ± 0.78
Hydrogel with active ingredients	74.20 ± 0.56	73.83 ± 0.73	73.94 ± 0.75	72.36 ± 0.67	72.50 ± 0.58
			a*		
Basic hydrogel	4.65 ± 0.47	3.99 ± 0.26	4.61 ± 0.53	4.64 ± 0.41	5.49 ± 0.24
Hydrogel with active ingredients	4.65 ± 0.47	4.34 ± 0.61	5.31 ± 0.18	5.11 ± 0.70	5.55 = 0.73
			ITA°		
Basic hydrogel	71.86 ± 0.80	71.63 ± 1.11	72.34 ± 0.45	71.90 ± 0.72	70.29 ± 0.36
Hydrogel with active ingredients	71.86 ± 0.80	71.95 ± 0.45	71.71 ± 0.81	71.86 ± 0.58	69.71 ± 0.84

19.3.2.2 SKIN HYDRATION

The interpretation of very dry, dry, and sufficiently hydrated values has been chosen according to the experience of the manufacturer of the Corneometer® CM825 probe and is only an approximate scale. Values are given in arbitrary units from 0 to 130, where values above 40 indicate sufficiently hydrated skin. The result of the analysis of skin hydration level after the application of polymer films with and without active ingredients is shown in Figure 19.5. After using both hydrogels, the skinFigures hydration level increased; however, the hydrogel with added active ingredients showed a much better moisturizing effect.

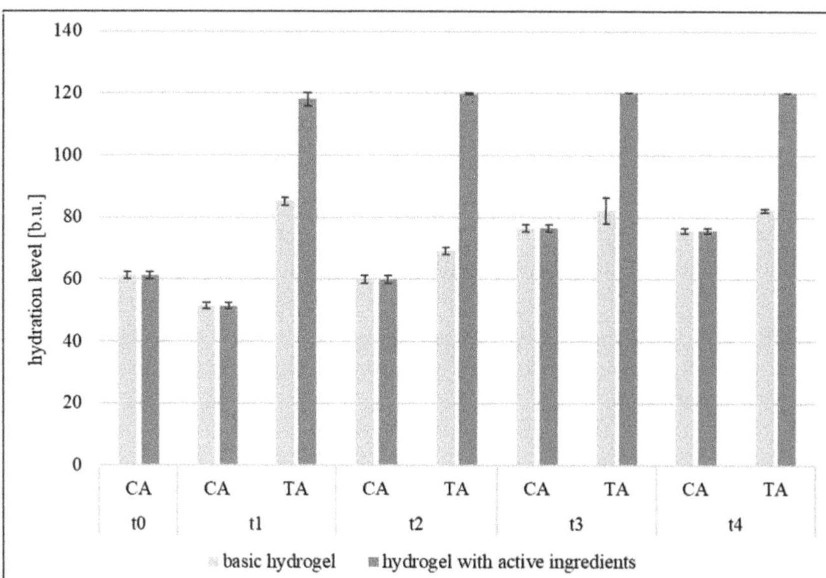

FIGURE 19.5 Skin hydration level after application of basic hydrogel and hydrogel with active ingredients (*Centella asiatica* extract, panthenol, royal jelly, and glycerol).

19.3.2.3 TRANSEPIDERMAL WATER LOSS (TEWL)

The result of skin parameter analysis after applying hydrogels with and without active ingredients is shown in Figure 19.6. The permeability barrier function was assessed as TEWL. The TEWL values were lower after hydrogel application. There has been an improvement in barrier function after application of both hydrogels. Nevertheless, better results were obtained after applying the basic hydrogel, possibly due to intense moisturizing effect of the active ingredients.

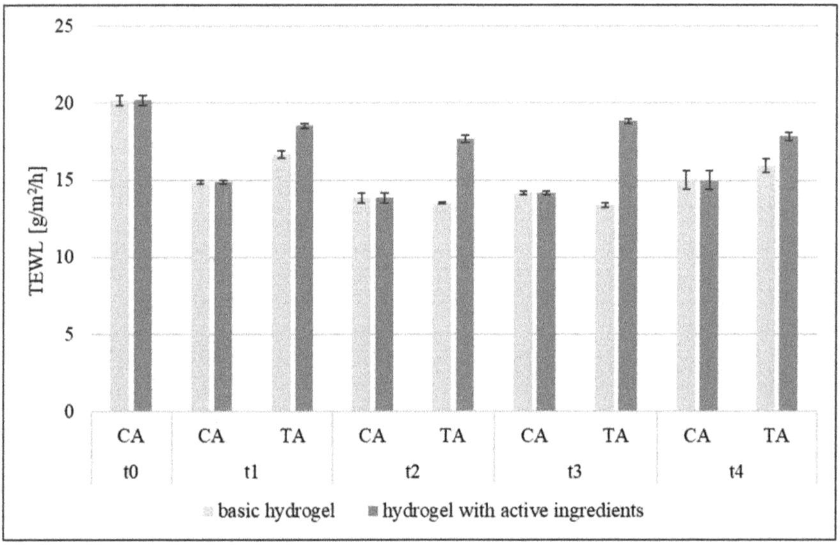

FIGURE 19.6 Skin barrier quality (transepidermal water loss - TEWL) after application of basic hydrogel and hydrogel with active ingredients (*Centella asiatica* extract, panthenol, royal jelly, and glycerol).

19.3.2.4 SKIN ELASTICITY

The device measures the degree to which the probe moves the skin. The pinFigures penetration depth (displacement) is measured in mm (0–3 mm). The firmer/stiffer the skin, the shallower the displacement made by the pin. The result of skin firmness analysis after applying polymer hydrogels with and without active ingredients is shown in Figure 19.7. Improvement in skin firmness was observed only immediately after removing the hydrogel with active ingredients. After a period of time, the skin returned to its original state. Perhaps long-term and regular use of hydrogels on the skin would improve its elasticity. Thus, a one-time hydrogel application is not enough.

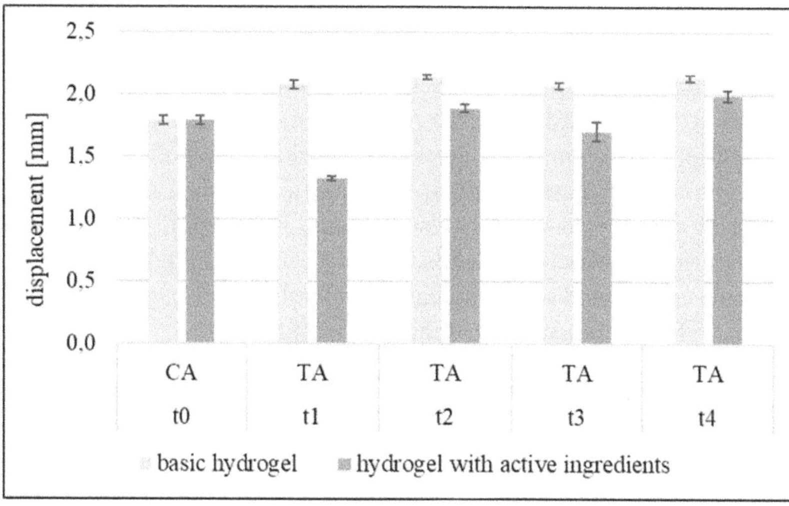

FIGURE 19.7 Skin firmness after application of basic hydrogel and hydrogel with active ingredients (*Centella asiatica* extract, panthenol, royal jelly, and glycerol).

19.3.2.5 SEBUMETER® SM 815

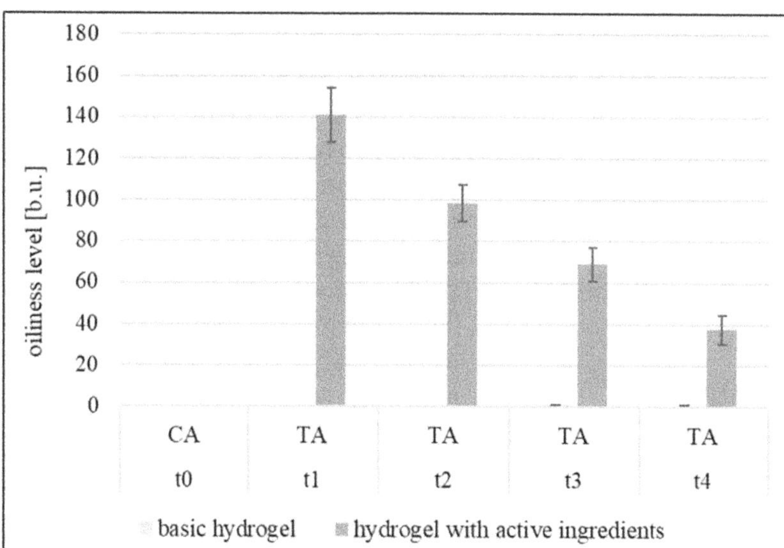

FIGURE 19.8 Skin oiliness after application of basic hydrogel and hydrogel with active ingredients (*Centella asiatica* extract, panthenol, royal jelly, and glycerol).

The result of the analysis of skin oiliness level after application of polymer hydrogels with and without active ingredients is shown in Figure 19.8. The basic hydrogel showed no lubricating effect. In contrast, the hydrogel with *C. asiatica* extract, panthenol, royal jelly, and glycerol increased the skinFigures oiliness level, which decreased over time. Besides proteins, sugars, phenols, flavonoids, vitamins, salts, minerals, and free amino acids, royal jelly also contains 3–6% lipids.[22] The forearm area is not rich in sebaceous glands, so the presence of lipids in royal jelly could have contributed to the increase in sebumetric measurements of this skin area, which is related to its lubricating effect.

19.4 CONCLUSION

Chitosan/sodium alginate-based hydrogels for skin applications were obtained successfully. In addition, we have shown that *C. asiatica* extract, panthenol, royal jelly, and glycerol may be loaded into such hydrogels during hydrogel preparation. The obtained hydrogels adhered very well to the skin. Additionally, adding any cross-linking agent was unnecessary because the swollen hydrogels were stable during application to the skin. The results of skin parameter tests after using the base hydrogel (without the addition of active ingredients) and the hydrogel enriched with active ingredients showed intensive moisturizing action, strengthening of the epidermal barrier, and no irritating effect. It should be emphasized that hydrogels with active ingredients additionally increased stratum corneum oiliness and moisturized the skin much more intensively. We believe the results of this work suggest that the obtained hydrogels may be helpful for several skin applications, such as cosmetic products, carriers of active substances, and dermatological preparations.

KEYWORDS

- **chitosan**
- **sodium alginate**
- **hydrogels**
- ***Centella asiatica extract***

- **royal jelly**
- **panthenol**
- **cosmetic applications**
- **dermatological applications**

REFERENCES

1. Barbieri, J. S.; Wanat, K.; Seykora, J. Skin: Basic Structure and Function. In *Pathobiology of Human Disease*; McManus, L. M., Mitchell, R. N., Eds.; Elsevier: Oxford, United Kingdom, 2014; pp 1134–1144.
2. Harding, C. R. The stratum corneum: Structure and Function in Health and Disease. *Dermatol. Ther*. **2004,** *17*, 6–15.
3. Gilaberte, Y.; Prieto-Torres, L.; Pastushenko, I.; Juarranz, Á. Anatomy and Function of the Skin. In *Nanoscience in Dermatology*; Hamblin, M. R., Avci, P., Prow, T., Eds.; Academic Press, 2016; pp 1–14.
4. Madhvapathy, S. R.; Arafa, H. M.; Patel, M.; Winograd, J.; Kong, J.; Zhu, J.; Xu, S.; Rogers, J. A. Advanced Thermal Sensing Techniques for Characterizing the Physical Properties of Skin. *Appl. Phys. Rev.* **2022,** *9 (4)*, 041307.
5. Gerhardt, L. C.; Strässle, V.; Lenz, A.; Spencer, N. D.; Derler, S. Influence of Epidermal Hydration on the Friction of Human Skin against Textiles. *J. R. Soc. Interface* **2008,** 51317–51328.
6. Pu, L.; Wang, H.; Zhao, Y.; Yuan, Z.; Zhang, Y.; Ding, J.; Qu, K.; Sun, W.; Xue, Z.; Xu, W.; Sun, X. Skin-Like Hydrogels: Design Strategy and Mechanism, Properties, and Sensing Applications. Skin-Like Hydrogels: Design Strategy and Mechanism, Properties, and Sensing Applications. *J. Mater. Chem. C* **2023,** *11*, 8358–8377.
7. Aranaz, I.; Alcántara, A. R.; Civera, M. C.; Arias, C.; Elorza, B.; Caballero, A. H.; Acosta, N. Chitosan: An Overview of Its Properties and Applications. *Polymers* **2021,** *13*, 3256.
8. Zargar, V.; Asghari, M.; Dashti, A. A Review on Chitin and Chitosan Polymers: Structure, Chemistry, Solubility, Derivatives, and Applications. *ChemBioEng Rev.* **2015,** *2*, 204–226.
9. Azmana, M.; Mahmood, S.; Hilles, A. R.; Rahman, A.; Arifin, M. A. B.; Ahmed, S.; A Review on Chitosan and Chitosan-Based Bionanocomposites: Promising Material for Combatting Global Issues and its Applications. *Int. J. Biol. Macromol.* **2021,** *185*, 832–848.
10. Pellis, A.; Guebitz, G. M.; Nyanhongo, G. S. Chitosan: Sources, Processing and Modification Techniques. *Gels* **2022,** *8 (7)*, 393.
11. Joseph, S. M.; Krishnamoorthy, S.; Paranthaman, R.; Moses, J.; Anandharamakrishnan, C. A Review on Source-Specific Chemistry, Functionality, and Applications of Chitin and Chitosan. *Carbohydr. Polym. Technol. Appl.* **2021,** *2*, 100036.

12. Jadach, B.; Świetlik, W.; Froelich, A. Sodium Alginate as a Pharmaceutical Excipient: Novel Applications of a Well-known Polymer. *J. Pharm. Sci.* **2022,** *5*, 1250–1261.
13. Lee, K. Y.; Mooney, D. J. Alginate: Properties and Biomedical Applications. *Prog. Polym. Sci.* **2012,** *37* (1), 106–126.
14. Abka-Khajouei, R.; Tounsi, L.; Shahabi, N.; Patel, A. K.; Abdelkafi, S.; Michaud, P. Structures, Properties and Applications of Alginates. *Mar. Drugs* **2022,** *20* (6), 364.
15. Dattilo, M.; Patitucci, F.; Prete, S.; Parisi, O. I.; Puoci, F. Polysaccharide-Based Hydrogels and Their Application as Drug Delivery Systems in Cancer Treatment: A Review. *J. Funct. Biomater.* **2023,** *14*, 55.
16. Sahoo, D. R.; Biswal, T. Alginate and its Application to Tissue Engineering. *SN Appl. Sci.* **2021,** *3*, 30.
17. Lee, K. Y.; Mooney, D. J. Alginate: Properties and Biomedical Applications. *Prog. Polym. Sci.* **2012,** *37* (1), 106–126.
18. Diniz, L. R. L.; Calado, L. L.; Duarte, A. B. S.; de Sousa, D. P. *Centella asiatica* and Its Metabolite Asiatic Acid: Wound Healing Effects and Therapeutic Potential. *Metabolites* **2023,** *13*, 276.
19. Ebner, F.; Heller, A.; Rippke, F. Topical Use of Dexpanthenol in Skin Disorders. *Am. J. Clin. Dermatol.* **2002,** *3*, 427–433.
20. Choudhary, V.; Kaddour-Djebbar, I.; Custer, V. E.; Uaratanawong, R.; Chen, X.; Cohen, E.; Yang, R.; Ajebo, E.; Hossack, S.; Bollag, W. B. Glycerol Improves Skin Lesion Development in the Imiquimod Mouse Model of Psoriasis: Experimental Confirmation of Anecdotal Reports from Patients with Psoriasis. *Int. J. Mol. Sci.* **2021,** *22* (16), 8749.
21. Kurek-Górecka, A.; Górecki, M.; Rzepecka-Stojko, A.; Balwierz, R.; Stojko, J. Bee Products in Dermatology and Skin Care. *Molecules* **2020,** *25* (3), 556.
22. Guo, J.; Wang, Z.; Chen, Y.; Cao, J.; Tian, W.; Ma, B.; Dong, Y. Active Components and Biological Functions of Royal Jelly. *J. Funct. Foods* **2021,** *82*, 104514.

CHAPTER 20

Machine Learning in Materials Science: Predicting Melt Flow Rate in Polymers

IRAKLI KOBERIDZE, MAGDA TSINTSADZE, and MANANA KHACHIDZE

Department of Computer Science, Tbilisi State University, Tbilisi, Georgia

ABSTRACT

This chapter presents a comprehensive study on predicting the melt flow rate (MFR) of polymers employing a variety of machine learning (ML) methods, including least absolute shrinkage and selection operator (LASSO) regression, random forest, XGBoost, artificial neural network (ANN), and support vector regressor. The study enriches the literature by incorporating various preprocessing tools for polymer data, such as modified *Z*-score, transformation of MFR to its natural logarithm, and standard scaling. As well as using grid-search and cross-validation to more comprehensively evaluate the algorithms. The performance of these models varied, with random forest and XGBoost showing the best performance and achieving an R^2 score of 0.8 on previously unseen data. The results underscore the significance of choosing the right model for ML tasks, showing that the complexity of the model and the type of data can lead to considerably different outcomes. This research contributes to advancing the field of polymer MFR prediction and encourages transparency and collaboration in scientific research. Furthermore, the study provides both the source code and the dataset used in the research to facilitate reproducibility.

Advances in Engineering Materials: Preparation for Sustainable Process Development. Tamara Tatrishvili and A. K. Haghi (Eds.)

20.1 INTRODUCTION

The field of materials science has witnessed significant advancements, necessitating the development of efficient and precise prediction models for polymer processing.[1] A critical parameter in this domain is the melt flow rate (MFR),[2] a determinant of the physical properties and applicability of polymers. The MFR quantifies the ease of flow of a thermoplastic polymer's melt, thereby influencing its processability in various applications, including injection molding, extrusion, and blow molding.

Predicting the MFR of polymers presents a complex challenge due to the intricate interplay between factors such as temperature, pressure, and molecular weight distribution.[3] Conventional prediction methodologies often rely on laborious experiments and complicated mathematical models, which may not yield consistently accurate results.[4,5]

Machine learning (ML), a subfield of Artificial Intelligence, is concerned with the development and implementation of algorithms that enable computers to learn from and make decisions or predictions based on data. In recent years, it has evolved into a potent instrument, demonstrating an exceptional ability to predict intricate material properties with remarkable precision and efficiency.[36]

ML methodologies, by virtue of their capacity to infer from extant data and project outcomes for novel data, exhibit a high degree of suitability for prognosticating the MFR of polymers. This ability to predict and analyze has found its application in various fields. For instance, in healthcare, ML can help in the early detection of diseases and personalized treatment. In finance, it can be used for credit scoring and algorithmic trading. In transportation, ML can optimize routes and schedules to improve efficiency.[37–39]

The continual evolution and expansion of ML applications in the realm of materials science, inclusive of polymer processing, is a testament to the growing recognition of its potential among the scientific community. This has catalyzed the rate of materials discovery, augmented the efficacy of polymer processing, and deepened our comprehension of materials at the atomic and molecular echelons. The amalgamation of computer science, data analysis, and domain-specific knowledge in an interdisciplinary manner is revolutionizing the landscape of materials science.[6]

A notable impediment in this field is the scarcity of publicly accessible datasets for research.[7] The absence of dataset availability in many studies hinders the reproducibility or enhancement of their results. This study confronts the existing challenge by leveraging a dataset that is publicly accessible via Kaggle and our implementation is openly shared on GitHub to guarantee reproducibility and to encourage additional scholarly exploration.

This study aims to develop an ML algorithm that can predict the MFR with high precision. The development of such a model is expected to substantially decrease the resources, both in terms of time and cost, typically required by conventional prediction techniques. This advancement could potentially lead to more streamlined processes in the field of polymer manufacturing.

20.2 LITERATURE REVIEW

20.2.1 PREVIOUS STUDIES ON PREDICTING POLYMER MFR

The prediction of polymer properties, such as the MFR, has been a topic of interest in many studies. Various methods have been employed to predict these properties, ranging from experimental studies (involving measuring MFR under different conditions of temperature, pressure, and molecular weight distribution) to computational models.[8] These empirical studies formed the foundation for understanding the influence of various factors on MFR. At the same time, mathematical models provide a theoretical framework for predicting MFR based on equations incorporating parameters such as viscosity, shear rate, and polymer characteristics.

A research endeavor proposed a novel model for chemical language processing, aiming to predict the glass transition temperature of polymers. This innovative model utilized the simplified molecular input line entry system (SMILES) strings of a polymer's repeating units as input data. The model exhibited commendable predictive performance and was subsequently employed for high-throughput screening on an unlabeled polymer database.[9] This underscores the variety of methodologies that have been used to predict polymer MFR.

20.2.2 USE OF MACHINE LEARNING IN MATERIALS SCIENCE AND SPECIFICALLY POLYMER PROCESSING

ML techniques started to be applied to materials science and polymer processing in the early 2000s, but the adoption was relatively limited. The widespread adoption of ML in materials science, including polymer processing, gained momentum in the 2010s. This was driven by advancements in ML algorithms, increased computational power, and the availability of large and diverse datasets related to material properties. ML has been employed to address a myriad of challenging problems, such as the development of efficient and predictive surrogate models for a range of material properties, screening and down-selection of novel candidate materials for targeted applications, and new methodologies to improve and further expedite molecular and atomistic simulations.[10]

Within the realm of polymer science, the application of ML methodologies has the potential to significantly diminish computational expenses, expedite the development process, and enhance the precision of computations. This has led to its emergence as one of the most promising investigative techniques in the screening of novel materials and the prediction of material properties.[11]

One notable study by Ledwani et al.[12] conducted a comparative analysis of prediction models for the MFR of C2 and C3 polymers synthesized using nanocatalysts. This study is an excellent example of how ML can be applied to predict polymer properties.[12]

Moreover, ML methods have been combined with coarse-grained molecular dynamics simulations to address existing challenges in polymer science at the molecular chain level. This unification has been utilized in three crucial areas: characterization of polymeric configurations, prediction of properties using a feed–forward approach, and reverse engineering design.[13]

While ML has shown great potential in advancing our understanding and ability to predict and optimize polymer properties, it is essential to note that while these techniques offer significant advantages, they also present new challenges that need to be addressed as we continue to refine them.

20.2.3 RESEARCH OPPORTUNITIES

While significant progress has been made in predicting the MFR of polymers and applying ML in materials science, our research seeks to explore the untapped potential and uncharted areas within this field.

First, while various predictive methods such as least absolute shrinkage and selection operator (LASSO) regression, random forests, XGBoost, artificial neural networks (ANN), and support vector machines/support vector regression (SVM/SVR) have been employed in previous studies, the use of a modified *Z*-score for data processing has not been extensively explored.[14] Our research incorporates this technique, providing a fresh perspective on data processing methods in predicting polymer MFR.

Second, we explore the less-examined area of transforming MFR to its natural logarithm and standard scaling. These transformations hold the potential to enhance the performance of predictive models, yet their application in the context of polymer MFR prediction remains relatively uncharted.[15]

Finally, reproducibility has been a significant issue in many studies as they often do not provide the code or dataset used.[16] This lack of transparency hinders further research and improvement of existing models. In contrast, our study makes the code and dataset readily available in the code availability section, encouraging other researchers to build upon our work and potentially enhance the predictive models.

By addressing the above points, our research contributes to advancing the field of polymer MFR prediction and encourages transparency and collaboration in scientific research.

20.3 METHODOLOGIES

20.3.1 DATASET DESCRIPTION

The raw dataset obtained from Kaggle[17] includes 2486 observations with the statistics as in the below table generated using DataFrame.[35]

	C3=	H2R	Pressure	Level	C2=	Cat	Temp	MFR
count	2486.000000	2560.000000	2484.000000	2484.000000	2484.00000	2486.000000	2484.000000	2564.000000
mean	25306.285729	0.178427	30.663706	77.651055	42525.14000	0.138530	80.144365	8.185218
std	1706.481672	0.077473	0.423345	0.919600	11331.86896	0.041869	0.823554	5.088696
min	16106.025000	0.000000	26.946344	74.575958	9610.46480	0.022162	77.760117	1.500000
25%	24361.632000	0.136000	30.446129	76.992151	34795.53500	0.113764	79.677458	3.700000
50%	25365.754500	0.173500	30.622631	77.494477	41550.56250	0.132986	80.044308	4.300000
75%	26398.452250	0.200000	30.925738	78.210867	50010.29500	0.156990	80.496296	12.900000
max	30083.688000	0.980000	32.674332	83.841675	106073.61000	0.677979	91.566544	38.000000

FIGURE 20.1 Cleaned dataframe.

Following are the features from the dataset and how they could potentially affect the MFR.[18]

Propylene (C3=) feed rate (kg/h) influences the polymerization process and the molecular weight of the resulting polymer, with a higher rate leading to a higher molecular weight and lower MFR, and vice versa.[19]

The hydrogen to C3= ratio can potentially influence the MFR by affecting the molecular weight and structure of the polymer. Changes in this ratio could alter the properties of the polymer, including its MFR.[20]

Reactor pressure significantly influences the MFR in a reactor. Higher pressure increases the reaction rate and alters the physical properties of reactants and products, impacting their flow rate and potentially requiring more robust equipment or altering pump efficiency. Safety considerations also come into play at higher pressures.[21]

The reactor bed level can significantly influence the MFR in reactor systems. An increase in bed level can increase the residence time, potentially decrease the MFR, and lead to greater contact between reactant particles, potentially increasing reaction rates.[25]

The flow rate of ethylene (C2=) can significantly influence the MFR in reactor systems. An increase in ethylene flow rate can decrease the residence time, potentially increase the MFR, and lead to greater contact between reactant particles, potentially increasing reaction rates.[22]

The Catalyst Feed Rate can significantly influence the MFR)in reactor systems. An increase in catalyst feed rate can decrease the residence time, potentially increase the MFR, and lead to greater contact between reactant particles, potentially increasing reaction rates.[23]

The temperature of a reactor can significantly influence the MFR in reactor systems. An increase in temperature can lead to more frequent successful collisions between reactant particles, potentially increasing reaction rates and thus influencing MFR.[24]

20.3.2 MACHINE LEARNING METHODS

There are several ML techniques and algorithms that can be used in the context of predicting polymer MFR or other materials science-related tasks. The choice of which ML method to use depends on factors such as the nature of the dataset, the specific characteristics of the problem, the availability of computational resources, and the desired level of interpretability. Below we are presenting some of these methods chosen for their specific strengths and suitability for addressing particular challenges associated with the prediction of polymer MFR. The selection is reflected by thoughtful consideration of the dataset's characteristics and our research goals, aiming to obtain accurate and interpretable results while managing challenges such as overfitting, nonlinearity, and high-dimensional data.

20.3.2.1 LASSO REGRESSION

The LASSO regression is a methodological approach in linear regression analysis that simultaneously conducts a variable selection and regularization. This dual functionality enhances the precision of predictions and the interpretability of the statistical model generated. The LASSO method incorporates a penalty equivalent to the absolute value of the magnitude of coefficients. This results in the minimization of the sum of the residuals augmented by the sum of the absolute value of the coefficients. The selection of this method for the present study was motivated by its capacity to manage high-dimensional data and its efficacy in feature selection.[26]

20.3.2.2 SUPPORT VECTOR REGRESSOR

The SVR is a support vector machine that uses the space between data points as a margin of error and predicts the most likely next point in a dataset. SVR has robust regularization parameters, which makes it effective in handling overfitting. However, it can be computationally intensive for larger datasets. In this study, SVR was used due to its effectiveness in handling complex, nonlinear relationships in data.[27]

20.3.2.3 ARTIFICIAL NEURAL NETWORK (ANN)

An ANN represents a computational paradigm that draws its inspiration from the processing mechanisms of biological neural networks present in the human cerebral cortex. It consists of interconnected processing nodes, or "neurons," and can learn complex patterns using a process called backpropagation. ANNs are particularly suitable for predicting polymer MFR due to their ability to model nonlinear relationships and their capacity for learning and improving over time.[28]

20.3.2.4 RANDOM FOREST

The Random Forest algorithm, an ensemble learning technique, functions by generating multiple decision trees during the training phase and delivering the average prediction of the individual trees. This method exhibits robustness against overfitting and is capable of efficiently managing large datasets with high dimensionality. Nevertheless, its performance may be compromised with sparse data or infrequent outcomes. In the context of this study, the Random Forest algorithm was selected due to its proficiency in handling data of high dimensionality and its resilience to overfitting.[29]

20.3.2.5 XGBOOST

XGBoost, an acronym for Extreme Gradient Boosting, is a meticulously optimized library for distributed gradient boosting. It is engineered with a focus on superior efficiency, adaptability, and portability. This library stands out due to its capacity to perform computations at an accelerated pace while maintaining precision, making it a valuable tool in the realm of ML. It produces an ensemble of weak prediction models, typically decision trees, and optimizes them using gradient descent optimization algorithms. XGBoost was chosen for this study due to its speed and performance, its ability to manage missing values, and its suitability for large datasets. [30]

20.3.3 EVALUATION METRICS

The performance of ML models in this study was evaluated using the following metrics, each with its specific characteristics and purposes:

The mean absolute error (MAE) is a statistical measure that quantifies the average magnitude of the absolute discrepancies between the predicted and actual observations. It is calculated by taking the absolute differences between these two sets of values and then computing their arithmetic mean. This metric assesses prediction accuracy in the context of regression models. It provides a straightforward measure of prediction error rates. Since it takes the absolute value of the differences, it only provides information on the magnitude of the error, not its direction.[31]

The root mean square error (RMSE) represents the square root of the mean value of the squared deviations between the predicted and actual observations. It is a quadratic scoring rule that quantifies the average magnitude of the error. The RMSE metric is more sensitive to more significant errors compared to the MAE, making it particularly useful in situations where more significant errors, especially, are undesirable. This sensitivity to larger errors provides a more stringent evaluation of the model's performance when accuracy is paramount.[31]

The coefficient of determination, commonly known as R-squared (R^2), is a statistical metric that quantifies the extent to which the variability in a dependent variable can be accounted for by the variability in one or more independent variables within a regression model. It serves as an index of model fit and measures the model's capacity to predict unseen samples accurately.[32]

The choice of these metrics is justified by the need for a comprehensive assessment of the model's performance. Each metric offers a unique lens through which the precision and efficacy of the model can be evaluated. MAE focuses on average error magnitude, RMSE emphasizes the impact of significant errors, and R^2 offers insights into the model's explanatory power. By using all three metrics together, the study ensures a well-rounded evaluation that considers different aspects of model performance and aids in making informed decisions about the model's suitability for the task at hand.

20.4 DATA ANALYSIS

20.4.1 PREPROCESSING THE DATA

The preprocessing of the data involved several steps to ensure that the data was clean, standardized, and ready for analysis.

First, any rows containing missing values were dropped. Next, the target variable MFR was transformed using a natural logarithm to ensure a normal distribution, and the data was standardized using the SkLearn StandardScaler.[34] This is important as many ML algorithms assume that the input variables have a Gaussian distribution.[33] Then, the outliers were removed using the Modified Z-score method.[14] This method calculates the Z-score based on the median absolute deviation rather than the standard deviation. This makes it more robust to outliers compared to the standard Z-score method. Finally, the dataset was split into training and testing sets using an 80:20 split. This allows us to evaluate the performance of our model on unseen data

These preprocessing steps help ensure that our ML models can be trained on clean, standardized data, improving their accuracy and reliability.

20.4.2 APPLICATION OF MACHINE LEARNING METHODS

20.4.2.1 LASSO REGRESSION

The LASSO regression model was applied to the preprocessed data. A grid search was executed to ascertain the most favorable value for the regularization parameter alpha. The grid search used 5-fold cross-validation and tested a range of potential alpha values from 1e-15 to 20. The best model, as determined by the grid search, was then used to make predictions on both the training and test sets. The performance of the model was evaluated using three metrics: root mean square error (RMSE), MAE, and R-squared (R^2). These metrics provide a comprehensive assessment of the model's performance, with RMSE indicating the average magnitude of the prediction error, MAE providing a measure of prediction error rates, and R^2 quantifying the proportion of the variance in a dependent variable that can be elucidated by one or more independent variables within a regression model is a statistical measure that signifies the proportion of the variance for a dependent variable that is explained by an independent variable or variables in a regression model. It essentially measures how well the unseen samples are likely to be predicted by the model based on the information it has been supplied with.

The best parameters for LASSO regression were found to be "alpha": 0.001, and the performance metrics for the training and test sets were as follows:

Training RMSE: 0.72, MAE: 0.60, R^2: 0.47
Test RMSE: 0.71, MAE: 0.6, R^2: 0.49

20.4.2.2 RANDOM FOREST

A grid search was performed to find the optimal values for the number of estimators (n_estimators) and the maximum depth of the tree (max_depth). The best parameters for random forest were found to be "max_depth": 40, "n_estimators": 80, and the performance metrics for the training and test sets were as follows:

Training RMSE: 0.17, MAE: 0.10, R^2: 0.96
Test RMSE: 0.48, MAE: 0.28, R^2: 0.77

20.4.2.3 XGBOOST

For XGBoos we found the optimal depth to be 3 with 320 estimators with these performance metrics:

Training RMSE: 0.28, MAE: 0.19, R^2: 0.91
Test RMSE: 0.44, MAE: 0.28, R^2: 0.8

20.4.2.4 ARTIFICIAL NEURAL NETWORK (ANN)

The ANN model was applied to the preprocessed data using PyTorch, a popular deep-learning library. The model architecture consisted of an input layer, two hidden layers with 64 and 32 neurons, respectively, and an output layer. The activation function used in the hidden layers was a rectified linear unit (ReLU). The computational model was subjected to a training process utilizing the mean square error (MSE) as the loss function. The optimization of this process was achieved through the implementation of the Adam optimization algorithm, set with a learning rate parameter of 0.001. A batch size of 16 was used for training. To prevent overfitting, early stopping was implemented. The training was stopped if the validation loss did not improve for 10 consecutive epochs. The best model state was saved whenever the validation loss improved.

Training RMSE: 0.17, MAE: 0.10, R^2: 0.97
Test RMSE: 0.48, MAE: 0.28, R^2: 0.77

20.4.2.5 SUPPORT VECTOR REGRESSOR

An optimal value for the kernel was determined to be the radial basis function (RBF), the penalty parameter C of the error term was 1, and an epsilon of 0.1. Yielding the following performance:

Training RMSE: 0.43, MAE: 0.28, R^2: 0.81
Test RMSE: 0.50, MAE: 0.33, R^2: 0.75

20.5 RESULTS AND DISCUSSION

Table 20.1 shows the summary of results obtained through experimentations.

TABLE 20.1 Prediction Accuracies of Various Models.

Method	RMSE (Train/Test)	MAE (Train/Test)	R^2 (Train/Test)
LASSO	0.72/0.71	0.6/0.6	0.47/0.49
Random Forest	0.17/0.48	0.10/0.28	0.96/0.77
XGBoost	0.28/0.44	0.19/0.28	0.91/0.8
ANN	0.30/0.47	0.21/0.29	0.89/0.76
SVR	0.43/0.50	0.28/0.33	0.81/0.75

The results presented in Table 20.1 provide a comparative analysis of the performance of five different ML models: LASSO, random forest, XGBoost, ANN, and SVR. The performance of these models is evaluated using three metrics: RMSE, MAE, and R^2 (coefficient of determination).

LASSO regression shows a consistent RMSE and MAE for training and testing datasets, indicating that the model is not overfitting. However, the R^2 values are relatively low, suggesting that this model may not capture all the complexities in the data.

Random forest exhibits the best performance on the training set with the lowest RMSE and MAE, and the highest R^2. However, its performance drops on the test set, which could indicate overfitting. This means that while the model performs exceptionally well on the training data, it may not generalize well to unseen data.

XGBoost and ANN show similar patterns in their performance. Both models have a moderate increase in RMSE and MAE from training to testing data, suggesting a good fit without overfitting. The R^2 values for both models are high, indicating that they can explain a significant proportion of the variance in the target variable.

SVR, on the other hand, has higher error rates (RMSE and MAE) compared to other models. Its R^2 values are also lower, suggesting that it may not be as effective in capturing the underlying patterns in the data.

The differences in performance among these models could be attributed to several factors. These include the nature of the data, the complexity of the model, the choice of hyperparameters, and how well each model can capture nonlinear relationships in the data. It is also important to note that while some models may perform better on specific metrics, they may not necessarily be the best choice depending on the specific requirements of a task. For instance, if interpretability is the key concern, simpler models such as LASSO might be preferred despite their lower performance. These results highlight the importance of model selection in ML tasks and demonstrate that different models can yield significantly different results depending on their complexity and the nature of the data.

20.6 CONCLUSION

The study applied five different ML methods—LASSO regression, random forest, XGBoost, ANN, and SVR—to predict the MFR of polymers. The performance of these models varied significantly.

LASSO regression had the highest RMSE values of 0.72 on the training set and 0.71 on the test set and the lowest R^2 scores of 0.47 and 0.49 on the training and test sets, respectively. This indicates that LASSO regression had lower prediction accuracy than other models.

Random Forest showed an impressive performance with the lowest RMSE of 0.17 on the training set and 0.48 on the test set. It also achieved the highest R^2 score of 0.96 on the training set, although it dropped to 0.77 on the test set.

XGBoost demonstrated strong performance with an RMSE of 0.28 on the training set and 0.44 on the test set and R^2 scores of 0.91 and 0.8 on the training and test sets, respectively.

The ANN model also performed well with an RMSE of 0.30 on the training set and 0.47 on the test set and R^2 scores of 0.89 and 0.76 on the training and test sets, respectively.

SVR had an RMSE of 0.43 on the training set and 0.50 on the test set and R^2 scores of 0.81 and 0.75 on the training and test sets, respectively, indicating a good balance between model complexity and prediction accuracy.

This research contributes to the field of materials science and polymer processing by introducing a novel approach to predicting polymer MFR using ML methods. It enriches the literature by incorporating a modified *Z*-score for data preprocessing, exploring the transformation of MFR to its natural logarithm and standard scaling while providing both the code and dataset for reproducibility.

The findings from this research have several implications for future research. Firstly, they highlight the importance of model selection in MFR prediction. Different models can yield significantly different results depending on their complexity and the nature of the data. Secondly, they demonstrate the effectiveness of data transformations and standardization in improving model performance. Finally, by providing both the code and dataset used in this study, this research encourages other researchers to build upon this work, potentially enhancing the predictive models.

Potential improvements to this research could include exploring other ML methods or tuning the parameters of the current models further to improve their performance. Additionally, future studies could investigate other data preprocessing techniques or transformations to see if they can further improve model performance.

CODE AVAILABILITY

https://github.com/irakli97/MFR_Prediction

KEYWORDS

- **polymer processing**
- **materials science**
- **polymer melt flow rate (MFR)**
- **machine learning**
- **predictive modeling**

REFERENCES

1. Peter, H.; Diamandis, M. D. 3 Major Materials Science Breakthroughs—and Why They Matter for the Future; May 21, 2020. https://singularityhub.com/2020/05/21/3-major-materials-science-breakthroughs-and-why-they-matter-for-the-future/ (accessed Oct 2023, 21).

2. Why is it Important to Test the Melt Flow Index of Polymers, [Online] Dec 14, 2022. https://www.pacorr.com/blog/important-to-test-the-melt-flow-index-of-polymers/ (accessed Oct 21, 2023)
3. Wilczyński, K.; Wilczyński, K. J.; Buziak, K. Modeling and Experimental Studies on Polymer Melting and Flow in Injection Molding. *Polymers* **2022,** *14*, 2106. https://doi.org/10.3390/polym14102106
4. Rides, M.; Allen, C. R. G. Guide to the Measurement of the Flow Properties of Polymers, [Online] Dec 2002. https://eprintspublications.npl.co.uk/2609/1/MATC_MN42.pdf (accessed Oct 21, 2023)
5. Dawson, P. C. Flow Properties of Molten Polymers. In *Mechanical Properties and Testing of Polymers, Polymer Science and Technology Series*; Swallowe, G. M., Eds.; Springer: Dordrecht, 1999; vol 3. https://doi.org/10.1007/978-94-015-9231-4_21
6. Rodrigues, J. F.; Florea, L.; de Oliveira, M. C. F.; et al. Big Data and Machine Learning for Materials Science. *Discov. Mater.* **2021,** *1*, 12. https://doi.org/10.1007/s43939-021-00012-0
7. Blaiszik, B.; Ward, L.; Schwarting, M.; et al. A Data Ecosystem to Support Machine Learning in Materials Science. *MRS Commun.* **2019,** *9*, 1125–1133. https://doi.org/10.1557/mrc.2019.118
8. Patra, T. K. *ACS Polymers Au* **2022,** *2* (1), 8–26. DOI: 10.1021/acspolymersau.1c00035
9. Chen, G.; Tao, L.; Li, Y. Predicting Polymers' Glass Transition Temperature by a Chemical Language Processing Model. *Polymers* **2021,** *13*, 1898. https://doi.org/10.3390/polym13111898
10. Li, Y. Q.; Jiang, Y.; Wang, L. Q.; et al. Data and Machine Learning in Polymer Science. *Chin J Polym. Sci.* **2023,** *41*, 1371–1376. https://doi.org/10.1007/s10118-022-2868-0
11. Huang, G.; Guo, Y.; Chen, Y.; Nie, Z. Application of Machine Learning in Material Synthesis and Property Prediction. *Materials* **2023,** *16*, 5977. https://doi.org/10.3390/ma16175977
12. Ledwani, D.; Thakur, I.; Bhatnagar, V. Comparative Analysis of Prediction Models for Melt Flow Rate of C2 and C3 Polymers Synthesized using Nanocatalysts. *NanoWorld J.* **2022,** *8* (S1), S123–S127.
13. Nguyen, D.; Tao, L.; Li, Y. Integration of Machine Learning and Coarse-Grained Molecular Simulations for Polymer Materials: Physical Understandings and Molecular Design. *Front. Chem.* **2022,** *9*, 820417. DOI: 10.3389/fchem.2021.820417
14. Wu, M.; Sun, M.; Zhang, F.; Wang, L.; Zhao, N.; Wang, J.; Huang, W. A Fault Detection Method of Electric Vehicle Battery Through Hausdorff Distance and Modified Z-Score for Real-World Data. *J. Energy Storage* **2023,** *60*, 106561. ISSN 2352-152X. https://doi.org/10.1016/j.est.2022.106561
15. Huilgol, P. Feature Transformation and Scaling Techniques to Boost Your Model Performance, [Online] Sept 15, 2023. https://www.analyticsvidhya.com/blog/2020/07/types-of-feature-transformation-and-scaling/ (accessed Oct 15, 2023)
16. Sohn, E. The Reproducibility Issues that Haunt Health-Care AI. *Nature* **2023,** *613*, 402–403. DOI: https://doi.org/10.1038/d41586-023-00023-2
17. Kaggle. *Predict MFR of Polymer*, [Online] 2019. https://www.kaggle.com/datasets/apek1999/predict-mpv-of-polymer/data (accessed Sept 15, 2023)

18. Apmonitor. *Polymer Melt Flow Rate*, [Online] 2022. https://apmonitor.com/pds/index.php/Main/PolymerMeltFlowRate (accessed Sept 15, 2023)
19. Polyolefin Flow Characteristics: More than Just Melt Index or Melt Flow Rate. [Online] 2023. https://www.lyondellbasell.com/globalassets/documents/polymers-technical-literature/tech-topic-polyolefin-flow-characteristics.pdf?id=13945 (accessed Sept 20, 2023)
20. Govindaswamy, P.; Wada, E.; Kono, H.; Uozumi, T.; Funabashi, H. Propylene Polymerization Performance with Ziegler-Natta Catalysts Combined with U-Donor and T01 Donor as External Donor. *Catalysts* **2022,** *12*, 864. https://doi.org/10.3390/catal12080864
21. Indika Senevirathna, M. K.; Gamage, S.; Atalay, R.; Acharya, A. R.; Unil Perera, A. G.; Dietz, N.; Buegler, M.; Hoffmann, A.; Su, L.; Melton, A.; Ferguson, I. Effect of Reactor Pressure on the Electrical and Structural Properties of InN Epilayers Grown by High-Pressure Chemical Vapor Deposition. *Vac. Sci. Technol. A* **2012,** *30*, 031511. https://doi.org/10.1116/1.4705727
22. Sun, B.; Dang, L.; Bi, Q.; Li, R.; Gong, Q.; Wan, Z.; Xu, S. Effect of Different Compatibilizers on the Mechanical, Flame Retardant, and Rheological Properties of Highly Filled Linear Low-Density Polyethylene/Magnesium Hydroxide Composites. *Polymers* **2023,** *15*, 4115. https://doi.org/10.3390/polym15204115
23. Rueangjitt, N.; Sreethawong, T.; Chavadej, S.; et al. Non-Oxidative Reforming of Methane in a Mini-Gliding Arc Discharge Reactor: Effects of Feed Methane Concentration, Feed Flow Rate, Electrode Gap Distance, Residence Time, and Catalyst Distance. *Plasma Chem Plasma Process* **2011,** *31*, 517–534. https://doi.org/10.1007/s11090-011-9299-y
24. Meixner, R.; Yildirim, F. A.; Schliewe, R. R.; Goebel, H.; Bauhofer, W.; Krautschneider, W. In *Low-Temperature Process for Manufacturing All Polymer Thin-Film Transistors*, Polytronic 2005-5th International Conference on Polymers and Adhesives in Microelectronics and Photonics, Warsaw, Poland, 2005; pp 197–199. DOI: 10.1109/POLYTR.2005.1596518.
25. Almukhtar, R. S.; Yahya, A. A.; Mahdy, O. S.; Majdi, H. S.; Mahdi, G. S.; Alwasiti, A. A.; Shnain, Z. Y.; Mohammadi, M.; AbdulRazak, A. A.; Philib, P.; et al. Numerical Analysis of Gas Hold-Up of Two-Phase Ebullated Bed Reactor. *ChemEngineering* **2023,** *7*, 101. https://doi.org/10.3390/chemengineering7050101
26. Kumar, D. *A Complete understanding of LASSO Regression*, May 30, 2023. https://www.mygreatlearning.com/blog/understanding-of-lasso-regression/ (accessed Oct 1, 2023)
27. Zhang, M.; Zhao, Y.; Yang, L. Robust Twin Support Vector Regression with Correntropy-Based Metric. *Multimed. Tools Appl.* **2023**. https://doi.org/10.1007/s11042-023-17315-4
28. Wong, K. K. L. Backpropagation Neural Network. In *Cybernetical Intelligence: Engineering Cybernetics with Machine Intelligence*; IEEE, 2024; pp 83–100. DOI: 10.1002/9781394217519.ch5.
29. Huang, J.; Chen, P.; Lu, L.; et al. WCDForest: A Weighted Cascade Deep Forest Model Toward the Classification Tasks. *Appl. Intell.* **2023**. https://doi.org/10.1007/s10489-023-04794-z

30. Latief, M. A.; Bustamam, A.; Siswantining, T. In *Performance Evaluation XGBoost in Handling Missing Value on Classification of Hepatocellular Carcinoma Gene Expression Data*, 2020 4th International Conference on Informatics and Computational Sciences (ICICoS), Semarang, Indonesia, 2020; pp 1–6. DOI: 10.1109/ICICoS51170.2020.9299012.
31. Willmott, C. J.; Matsuura, K. Advantages of the Mean Absolute Error (MAE) over the Root Mean Square Error (RMSE) in Assessing Average Model Performance. *Clim. Res.* **2005,** *30* (1), 79–82.
32. Nagelkerke, N. J. D. A Note on a General Definition of the Coefficient of Determination. *Biometrika* **1991,** *78*, 691–692.
33. Huang, H.; Liao, Z.; Wei, X.; Zhou, Y. Combined Gaussian Mixture Model and Pathfinder Algorithm for Data Clustering. *Entropy* **2023,** *25*, 946. https://doi.org/10.3390/e25060946
34. Buitinck, L.; Louppe, G.; Blondel, M.; Pedregosa, F.; Mueller, A.; Grisel, O.; Niculae, V.; Prettenhofer, P.; Gramfort, A.; Grobler, J.; Layton, R.; VanderPlas, J.; Joly, A.; Holt, B.; Varoquaux, G. In *API Design for Machine Learning Software: Experiences from the Scikit-Learn Project*, ECML PKDD Workshop: Languages for Data Mining and Machine Learning, 2013; pp 108–122.
35. McKinney, W. In *Data Structures for Statistical Computing in Python*, Proceedings of the 9th Python in Science Conference, 2010; pp 51–56.
36. Mitchell, T. M. Machine Learning; McGraw-Hill: New York, 1997; vol 1.
37. Rajkomar, A.; Dean, J.; Kohane, I. Machine Learning in Medicine. *New Engl. J. Med.* **2019,** *380* (14), 1347–1358.
38. Lo, A. W. Machine Learning Applications: What We Can Learn from Healthcare Finance. *CFA Inst. Res. Found. Briefs* **2020,** *5* (2), 1–15.
39. Vlahogianni, E. I.; Golias, J. C.; Karlaftis, M. G. Short-Term Traffic Forecasting: Where We are and Where We're Going. *Transp. Res. C Emerg. Technol.* **2014,** *43*, 3–19.

Index

D

E

F

For Product Safety Concerns and Information please contact our EU
representative GPSR@taylorandfrancis.com
Taylor & Francis Verlag GmbH, Kaufingerstraße 24, 80331 München, Germany

www.ingramcontent.com/pod-product-compliance
Ingram Content Group UK Ltd.
Pitfield, Milton Keynes, MK11 3LW, UK
UKHW021502240926
475702UK00035B/209

* 9 7 8 1 7 7 9 6 4 3 7 2 8 *